Animals
in the
Night

Books by J. H. Prince

ANIMALS IN THE NIGHT
ANATOMY AND HISTOLOGY OF THE EYE AND ORBIT
    IN DOMESTIC ANIMALS *(editor)*
COMPARATIVE ANATOMY OF THE EYE
THE RABBIT IN EYE RESEARCH
VISUAL DEVELOPMENT
YOU AND YOUR EYES

Animals in the Night

Senses in Action After Dark

by

J. H. PRINCE

THOMAS NELSON INC.

New York / Camden

Library of Congress Catalog Card Number: 71–145917
ISBN 0–8407–6142–2

Color photos on jacket by Allan Roberts

Contents

Animals in the Night

Some words used in describing animal life

The more than 900,000 species in the animal kingdom are divided into two main subkingdoms: those animals with backbones, called *vertebrates*, and those without backbones, called *invertebrates*, the *in* being a Latin prefix meaning "without."

The invertebrates include insects, crustaceans (crabs, crayfish, etc.), shellfish, and cephalopods (octopuses, squid, cuttlefish). The word *cephalopod* is made up of two Greek words, *kephale*, meaning "a head," and *podos*, meaning "a foot"; literally, "feet on the head," and this is just about the best description of an octopus.

When we speak of *carnivores*, we mean animals that eat flesh, like the big cats; the Latin word *carnis* means "flesh," and *voro* means "to devour." *Herbivores* eat vegetation, and that needs no explanation. We ourselves are *omnivorous*, for we eat flesh and vegetation.

Rodents can chew their way through matter such as wood; the word comes from the Latin *rodo*, meaning "to gnaw." *Ungulates*, from the Latin *ungula*, are animals that have hooves, which are really extended toes; *ruminants* chew cud, from the Latin word *rumino*, meaning "to chew again."

Amphibians spend their early life in water, and their later adult life mainly on land (frogs, salamanders, newts). The word is derived from the Greek *amphibios*, meaning "to lead a double life." *Teleost* fishes are those with skeletons of bone, from the Greek words *teleos,* meaning "perfect" or "normal," and *osteon,* "bone." *Elasmobranchs* (sharks and rays) have skeletons of cartilage (a softer material than bone). The Greek word *elasmos* means "a thin plate," and this is how much of the skeleton appears. *Branchia* means "a gill" in Latin. The word *reptile* comes from the Latin word for "creeping." When an animal is said to be *predatory,* it is a killer, from the Latin word that means "a plunderer."

The warm-blooded animals we call *mammals* (from the Latin *mamma,* which means "having breasts") are divided into three main groups. *Monotremes* (platypus and anteater) lay eggs but suckle their young. The Greek word *monos* means "single," and *trema,* "a hole"; the female excretory and sexual channels have one common outlet. *Marsupials* keep and suckle their young in a pouch. The Greek word *marsipos* means a "purse" or "bag."

The largest group of mammals, *placentals,* bear their young live and suckle them, but have no pouch. The Latin word *placenta* and the Greek word *plakentos* mean a "small round cake." The placenta is the nourishing membrane which is expelled when or after the offspring leaves the mother's womb.

The animal world
comes to life at night

When men settled their differences by fighting hand to hand, they needed only eyes and ears to know where the enemy was, and what he was doing. But as inventions began to bring about contact with enemies from distances beyond hearing and visual range, men had to produce special devices for detection, observation, and evasion, to say nothing of communication. Range finders, telescopes, radio, radar, sonar, television, light beams, and infrared sensors are some of these. But few such devices were really invented by man. The animal world had most of them long ago.

HUNTING AND SURVIVING IN THE DARK

Animals use sonar, radar, telescopes, range finders, light beams, and infrared sensors to hunt, and to avoid being killed, in the hours of darkness; but we are not aware of this when we merely see them around in the daytime. As we become more familiar with their nocturnal activities, however, and observe these devices in use, we understand better why so many animals are much more efficient than we are at hunting in the dark.

A motorist driving along a quiet country road at night will often see a pair of brightly shining eyes reflecting the light from his headlights; but if he passes that same place during daylight hours, the chances are that he will see nothing but the under-

11

1. The raccoon
(*Procyon lotor*), although
arhythmic, is more active
at night, when it forages
and brings its young
out of hiding.

growth. Anyone living near a swamp, a pond, or a lake is conscious of unusual sounds of activity in the hours of darkness that never seem to be heard in the daytime. These may be only the croaking of frogs, or the splashing of an alligator, but they will certainly seem noisier at night.

People who live close to a zoo will be even more aware of this. A zoo really comes to life after darkness falls. Eyes that during the day had been gold or brown, with fine dark slits for pupils, become large green or red shining orbs, and they move disturbingly from one spot to another like twin circles of animated radium suspended in midair. Rustlings, coughing, snorts, and a host of other startling sounds produce an uncomfortable atmosphere of restless activity as the animals respond to their instinct to be on the prowl.

Most of us have heard tales of someone going to bed in the tropics or subtropics and waking in the morning to discover that a snake had found its way in during the night, and had coiled up somewhere in the room. The trapper will tell us that his traps are more likely to be filled at night than during the day, no matter what part of the world he may be in. Such tales confirm that a very large part of the animal world goes about its necessary excursions and adventures under cover of darkness.

Everyone knows that the owl and the bat are active only at night. These two examples are given to us at such an early age that they come to mind first when thinking of night activity, and we overlook the fact that most other living creatures hide away in the daylight hours too. In fact, the majority of wild animals are either entirely nocturnal, or what is called *arhythmic* (without rhythm), which means that they can be as active at night as during the day, for nature has endowed them with senses that will function equally well under both conditions.

It is difficult to calculate exact numbers, but if we consider only the mammals, we can at least say that of the 4,237 known species, all 981 species of bats are nocturnal; perhaps 80 percent of the approximately 250 species of marsupials are either nocturnal or arhythmic, as are 60 percent of the 252 species of carnivores (lions, tigers, etc.), 30 percent of 210 species of ungulates (zebras, goats, cattle, etc.), 40 percent of the 1,729 species of rodents, and possibly 20 percent of the 193 species of primates (monkeys, lemurs, apes, etc.).

Based on these figures, a rough guess suggests that 85 percent of all mammals are either nocturnal or active at night some of the time. If a similar percentage applies to all other kinds of animals, it is obvious that we see only a very small part of the active animal world if we do not go looking specially. It is true that most birds are active only during the daytime, and so are a

2. Bears are also arhythmic, but unlike many other arhythmic animals, they do not seem to have any preference for night prowling. Most of them can be seen going about their normal activities during daylight.

3. The big cats often sleep through the warm day and start their activity at night. But there is always one alert sentry, its sensitive eyes protected by pupils that contract down to tiny circles. In this these cats are different from the domestic cat and some of the smaller wild ones, which have vertically slit pupils.

number of lizards, a few snakes and tortoises, and a few fish and amphibians.

Arhythmic mammals, which are active both by day and night, are all around us: the dog and its relatives the wolf, dingo, and fox; the cattle and their relatives the buffalo, the wild goat, the wild horse, the zebra; the cats and their relatives the lion, the tiger, the jaguar, the puma, the leopard, and so on. All these are active at night. Even bears and raccoons, which may be seen during the day, are also active at night, the raccoons especially; it is likely that most of the animal world's mating is carried out at night.

To develop senses that function equally well by day and by night was no simple feat. For instance, if we were able to see at night as well as the cat, we would have very much more sensitive eyes than we have now. This means that during the day, especially in bright sunlight, we would need special protective devices to prevent these extra-sensitive eyes from being damaged by excessive light. Nature has taken care of such problems for the nocturnal animals at the same time as it has endowed them with a remarkable ability to detect, sense, or realize the presence of other animals or things about them in complete darkness.

The bat, we know, is able to detect objects by sending out a continuous high-pitched pulsing sound, too high for the human ear to register, which is reflected by obstructions in its path, or by other bats or by insects, and picked up again as echoes by the bat's ears; the intensity of the echoes and their direction tell the bat instantaneously what is happening.

4. The coconut crab (*Birgus latro*) is wholly adapted to a land life and never goes near water. On land, however, it has many enemies, and so has become entirely nocturnal, like so many other animals. During the day it stays down in its burrow (top picture), but at night it ventures out and climbs the trees for coconuts (bottom picture).

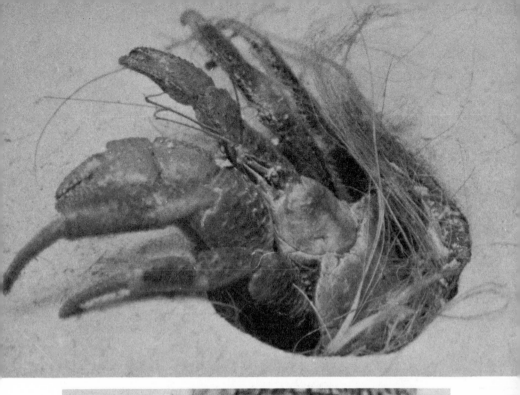

5. This lizard has completely sacrificed its legs, and moves like a snake. *Lialis burtonis* hunts for insects and other small animals at night, and its vertically slit pupils, which can close almost completely, emphasize the need to protect very sensitive retinas in the light. Unlike snakes, it also has ear openings (*E*), which some snake-lizards have discarded, suggesting it has retained external hearing.

Some fish have organs that will register the tiniest electrical currents produced by the movements of other fishes' gills, and can detect the minutest changes of water current produced by another fish moving at even a considerable distance. There are reptiles that have special organs for smelling or tasting the air, much more effective than the nose of a bloodhound or any other animal, and with these they can track down prey they have bitten, but which has escaped in the dark to die. These are perhaps the most sensitive tracking organs in the animal world.

Another device that is designed for use in darkness is found in certain snakes that are sensitive to the minutest change of temperature, fractions of a degree, in fact; and to infrared rays which will become noticeable when any warm-blooded creature passes by. Such sophisticated devices are a far cry from the cat's whiskers or the fish's barbels, which are also special sense organs.

Why are animals active at night? And why are so many of them secretive during the day? This is quite an interesting story.

AVOIDING THE LIGHT

As early animals grew more complex and their sense organs became fairly efficient, they began to swim around much more actively and prey upon one another more successfully. Some found it was easier to survive if they did not venture into the open while the light was bright enough for them to be seen and hunted. However, it is one thing to have a need to be active only in the dark, and quite another thing to be able to.

Gradually, many of the animals that were hunted most developed more sensitive eyes, and so were able to feed in dimmer light. This enabled them to stay hidden for more of the bright daylight period, so they also survived longer and consequently had a longer breeding span in which to pass on their characteristics.

The hunting animals were compelled to develop more sensi-tive eyes to capture their prey in the less well-lighted hours as well as in the bright daylight. Initially, the periods of increased activity may have been the twilight hours, around sunset and sunrise, but there were other animals that moved down to deeper zones in the sea during the daylight hours. Light does not penetrate water to the same extent it does air, so it is not necessary to go very far down to find a twilight zone in the waters of many parts of the world, even at midday.

By studying the eyes of nocturnal animals and analyzing the nerve endings, we learn a great deal about their history as well as their present habits. We know that when animals like the octopus and squid began to spread in large numbers, they had already become prey for many of the predatory sea inhabitants, especially the backboned and armor-plated fish. Not the least of these predatory types were the early forms of sharks.

The cephalopods were able to develop very sensitive eyes, and they learned to survive by hiding in dim places in the daytime and being active mainly in the hours of darkness. The sharks, too, developed exceedingly sensitive eyes, for whatever the hunted animal did in the way of changing its habits, to that, in order to thrive, the hunter had to adapt itself.

What happened to the cephalopods happened to many kinds of fish, especially if these were smaller than the prevalent preda-tors. Many gradually went deeper in the oceans, where light seldom penetrates even in a glimmer, until the far depths became fairly well populated with animals trying to escape predators.

The development of special sense organs for detecting prey in the dark has been part of an endless effort in the animal world to survive—by finding food, and by avoiding becoming another animal's food. This is more likely to be successful in darkness

6. The three-toed skink (*Siaphos equalus*) is a snake-lizard that has not yet lost its legs entirely. They have degenerated to mere vestiges, however, and are quite useless. This skink is secretive and hides under logs and leaves; it burrows rapidly, but it is not completely nocturnal. (Bottom picture): *Pygopus lepidopodus,* another snake-lizard, has retained only mere skin flaps (*L*) from its legs. It too hides under rocks, etc., but at times climbs bushes. Like *Siaphos,* it is arhythmic, but unlike *Siaphos,* it has retained openings for its ears (*E*).

than in light, but a life in darkness needs extreme efficiency for survival.

At a later period, land animals encountered the same problems as those that had been met and overcome by animals living in water. As soon as population growth made the land less comfortable because of the increasing quest for food, one of the first things some animals did to escape from powerful enemies was to burrow underground. Others leaped into and out of trees when pursued, and this eventually led to flying, by first gliding out of the trees and off high places.

Most animals that could do neither adopted a nocturnal life. Even the invertebrates did this and, as land animals became more specialized and more expert at hiding, new methods for detecting prey as well as for escaping hunters had to be devised. Ears became more sensitive; so did noses. Special organs were developed such as the infrared sense organs of snakes and the sound transmission equipment of bats and the mammals that, like the dolphin, adopted life in the oceans. It is possible that we still do not know all that nature has done to make animals aware, in some ways more aware than we are, of what is going on around them.

Perhaps the simplest example of this can be experienced by someone living in an old house infested with mice. When the

7. The green tree python (*Python viridis*) rests during the day, its color merging with the foliage, and its sensitive eyes protected by almost completely closed pupils, even though it is well shaded. At night it searches out birds and snaps up small nocturnal mammals. In the Americas, the green tree boa, or emerald boa, is virtually identical in habits and coloring with this snake. (Bottom picture): The blind worm-snake (*Typhlops polygrammicus*) is a true burrower, one of more than a hundred species that have sacrificed their eyes almost completely. These have degenerated to mere pigmented spots, which are able only to distinguish between light and dark. This animal comes above ground at night, following its food—earthworms, ants, and termites.

mice come out searching for food at night he may hear them, may even have a fair idea of the direction from which the sounds are coming, but he will not see them as a cat will. He will not be able to move his ears directionally to pick up the exact position of the sounds within an inch as a cat can, and thus that cat is more aware than he can be of what is happening. This can be expanded in many directions to show that other animals are clearly aware of what is going on around them in the dark, when we perhaps are merely bewildered by a host of different impressions.

It is very interesting that there are so few nocturnal birds, and almost all those that are nocturnal, such as the owl and the kiwi, have this habit because their food is nocturnal also. The owl hunts mice and small rodents; the kiwi hunts worms. Most other birds live on grain, insects, berries, and animals such as small lizards and other birds that are active in the daytime. Birds

8. The cave python is a powerful snake that has never entirely given up its secretive existence. It has sensitive eyes, which give it an advantage over other animals that may wander into its cave out of the bright light; and it can overpower most of them.

as a whole do not need to be nocturnal like so many other animals, because their general habit of perching, nesting, and living in trees and inaccessible places enables them to retreat to the safety these provide when there is danger.

Although man and the animals closely related to him are predominantly active during the day, they are nevertheless capable of some nocturnal activity. Man has no problems on a moonlit night, but he does on a really dark night, and so he is not ordinarily active at this time outside the well-lighted cities, either from preference or need. Once, his ancestors had to be as alert at night as during the day, not because they hunted then, but because they could be hunted by the larger animals. For a very long time now, man has been active mostly in the daylight hours, and the sensitivity of his eyes, ears, and nose has consequently become blunted.

RECOGNITION IN THE DARK

Other interesting devices are those that will enable animals to recognize their own species in the dark. We see countless forms of deep-sea fish with patterns of luminous organs that are the same throughout a given species, with slight differences in the two sexes, so that a male may recognize a female, or the other way around. These patterns of lights designed for recognition in the dark are like colors that are the same throughout a species of fish active in the daytime, again with differences in the sexes. Much the same can be said of birds; the plumage is unvaried for all females of a species, and the different plumage of the males also does not vary. The brilliant-red American male cardinal has a brown mate; so has the British blackbird, and the Australian blue wren.

PATTERNS OF BURROWING

There are two methods of burrowing away from the light. In one, the animal uses a hole which it digs, as the badger does, or a cave, and comes out mostly when the light has faded; in the other, the animal actually digs itself in, like the mole or the gopher. Both patterns are found underwater, on land, and even a few birds do one or the other. Many early snakes were true burrowers. Several snake species still live in caves; and almost all those which hibernate in winter find a hiding place in a dark, inaccessible hole.

Snakes were once lizards that gave up their legs for a new kind of locomotion, and for easier burrowing. There are still nocturnal lizards that are going through a similar change right now. Some have mere vestiges of legs, and others have even discarded these. They are all equipped with sensitive eyes, but, unlike true snakes, they still have ears, and they are still more lizard than snake, even though some burrow right down into the soft soil and sand like worms. These snake-lizards show the course of development once taken by snakes; they glide just like snakes, and they survive only in Australasia.

Few snakes burrow or live in caves now. The pressures on them have eased, and their venom has become effective against almost all their enemies, or, as with the pythons, they have become powerful enough to roam more freely. A few have moved up into the trees; but some never returned to an aboveground life. These have degenerated into a completely fossorial (digging) existence, and the few that we know show that their eyes have become almost useless and their ears are degenerate; they could never survive at all in the world above. There are over a hundred species of these burrowing snakes, and they are found right around the world—from the southern United States through tropical America, the West Indies, southern Europe, Africa, Asia, and Malaysia to Australia.

Seeing in the dark

Although the speed of an animal's movements, its running, swimming, or flying ability, enables it to escape from its hunters, or to catch what it hunts, undoubtedly the most important survival features in most animals are the warning senses of hearing, vision, and smell; this applies especially to nocturnal animals, which must use them to the very limit. When we examine these senses in the various animals, therefore, we get a clear picture of their way of life.

Even in the faintest light, eyes are used for the recognition of movement, contrast, shape, distance, and position. When there is more light, they also register shadow, color, and brightness. With the aid of memory, all this enables an animal to decide its whereabouts, and the closeness or distance of any visible danger. The other senses take over if there is an obstruction to vision. Hearing and smell can also detect danger and at times even judge its distance, but even these senses can be obstructed to some extent by wind and other noises.

Sensitive Eyes

The eye is quite a remarkable organ; in simple terms it is somewhat like a television camera, because the light that is received inside it can be transmitted along nerves to the brain just as an image can be transmitted along a television camera's

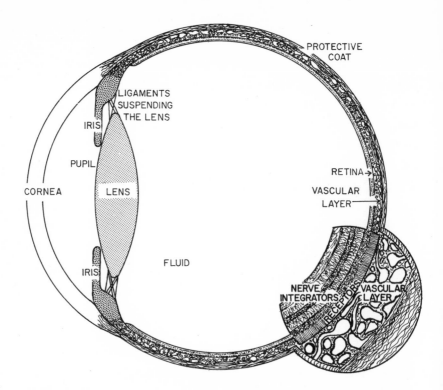

PROTECTIVE
COAT

LIGAMENTS
SUSPENDING
THE LENS

IRIS

PUPIL

RETINA→

CORNEA   LENS

VASCULAR
LAYER

IRIS

FLUID

NERVE   VASCULAR
INTEGRATORS   LAYER

RECEPTORS

9. If a normal eye is cut through from front to back, it is seen that the transparent cornea, or window, covers the colored iris in which is the pupil hole, and light passes through this to the interior of the eye. Variations in the size of the pupil control the intensity of light permitted through. Behind it is the lens that focuses the light onto the sensitive retina from which nerve fibers pass to the brain to produce the picture seen. Behind the retina is a complex vascular (blood-vessel) layer that nourishes the eye, and outside this is a tough layer that, together with the fluid filling the eye, keeps the eye in shape. A magnified inset shows that, as well as the receiving cells, the retina consists of many layers of other cells and fibers, which are concerned with sorting out, amplifying, and integrating parts of the image formed in the eye.

cables to the transmitter or receiver it serves. This description of an eye may be adequate to explain the simple principle, but at the microscopic level the whole apparatus is so complicated that it will be some time before we have unraveled the entire circuitry of the minute nerve systems. It is a more complex arrangement than any electronic system devised by man so far.

We need to concern ourselves with only a few of the microscopic details here. For example, there are two kinds of light receptor cells in the retina of the eye, each with its own minute nerve ending, and packed thousands to a square millimeter. One is called a *cone*, because of its shape, and this functions in bright light. The other, called a *rod*, again because of its shape, is found in much greater numbers in most animals and is much more sensitive, so it is the one used in very low illumination and near-darkness. It is the rods that become highly specialized in the animals that are active at night, and frequently their extreme sensitivity is amplified or aided by other special devices, such as a

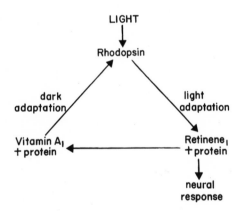

10. The action of light on the chemical rhodopsin in the retinal receptors of the eye: light breaks down the chemical to stimulate nervous impulses to the brain; a protein in the residual material then combines with vitamin A to remanufacture the rhodopsin. The process of breaking down is known as light adaptation, and that of building up as dark adaptation.

pupil that opens widely in low illumination to allow the maximum amount of light into the eye.

When light hits the rods, some of it is absorbed by an exceedingly sensitive chemical in them called *rhodopsin*, which then bleaches and breaks down into two other chemicals. One of these is called *retinene*, and the other is a *protein*. This breaking-down action sets up electrical impulses to the brain, and the sum of all the impulses from all the rods in an eye receiving an image is what makes up the dim-light picture in the brain.

Any chemical that is broken down must be instantly built up again, otherwise function ceases. Rhodopsin is built up again with vitamin A, delivered constantly by the blood vessels behind the retina, and which lie adjacent to the rods and cones. The protein combines with this vitamin A to resynthesize or remanufacture rhodopsin. This breaking down and building up of rhodopsin is continuous and perfectly balanced, unless the eyes are subjected to glare.

Glare reduces the efficiency of the retina to a point where it takes such a long time to gather enough vitamin A for proper functioning that a temporary blinding is the result, and this can last for many hours if it is severe. In fact, there is evidence that some nocturnal animals may suffer permanent damage to their retinas by overexposure to light.

When we turn out the light in a room, or move from a lighted room into the dark, we require a short time for our eyes to become adapted to the dimmer conditions because the rod receptors that have been glared in the light need to take in greater quantities of vitamin A to build up the chemical for dim-light vision and take over from the cones that were used in the light. The younger a person is, the quicker this takes place, unless the vitamin A has been destroyed or neutralized in the blood by

11. The coyote uses its eyes when there is enough light, but at night, when there is no moon, its nose functions with its ears to give it all the information it needs about its surroundings.

excess alcohol, nicotine, or disease. This short period of adaptation is necessary in some animals also.

When a cat is put out at night, it seldom moves from just outside the door until it has become adapted to the dimness enough to discern outlines and movements. A dog, on the other hand, may merely shake itself and trot off into the night, satisfied that all the information it needs will reach it through its nose and ears; it can take its time to become adapted to the dimness. Both animals are doing what their ancestors did for countless generations in the wild state, using their most important senses, the cat its eyes, and the dog its nose.

There are two essential factors involved in seeing at night: protection of the sensitive rods during the day and a good supply of vitamin A. The liver is a storehouse for vitamin A, and it is interesting that some nocturnal carnivorous animals seek out the liver of their prey before eating anything else.

Not all retinal receptors are the same size. Cones are nearly always larger than rods, but either can be different sizes in different species of animal; and this is important. When a newspaper illustration is viewed through a magnifying glass, it is seen to

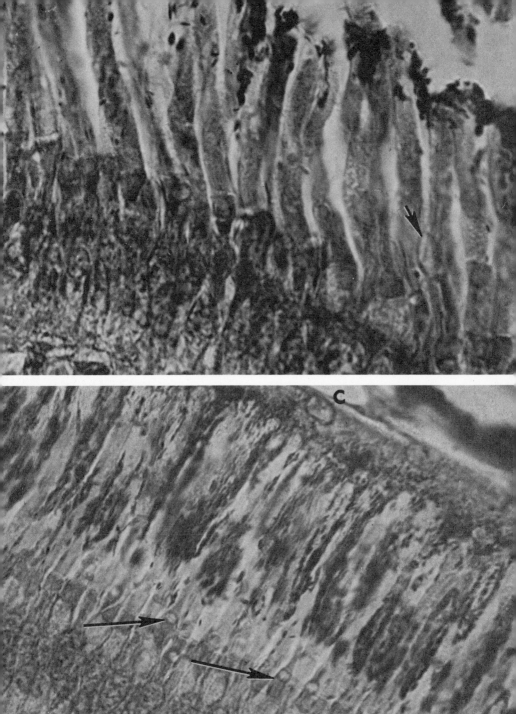

consist of countless dots. If the dots are very small, the picture shows great detail, but if the dots are coarse, then the picture shows far less detail. It is the same with eyes. If the receptors are small, packed many thousands to a square millimeter, then the image they form in the brain is very clearly defined and detailed; but coarse receptors, especially if they are rods, produce a very poor image.

These have one advantage, however. Being large, they contain more rhodopsin, and this makes them more sensitive in dimmer light, but it does not enable them to record a finely detailed

13. Small, sensitive receptors produce a finely detailed, clear image that will make it possible to recognize shape and markings even in dim light (left). Large receptors, on the other hand, no matter how sensitive, will not produce a detailed image, and anything seen at night will be more like a silhouette (right).

12. These two photographs, taken through a high-powered microscope, are from the area called receptors in Figure 9. (Top): The sensitive rods, used at night, can be clearly seen with some dense pigment at their tips. The arrow points to a single-day vision cone. (Lower): The pigment has migrated down among the rods to protect them from light, while a few cones (two are indicated by arrows) have withdrawn from the pigment zone to function in the light. C indicates one of the cells from which the pigment has migrated.

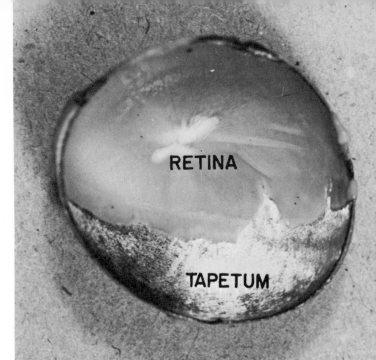

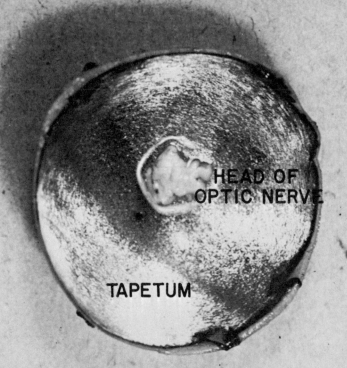

image. We find receptor populations in the animal world ranging
from 10,000 to 1 million to the square millimeter of retina, so the
range of quality in vision is quite extreme. But we can be sure
that any particular animal has evolved the size and type of receptor best suited to its way of life.

AMPLIFYING DIM LIGHT

One device for increasing the effectiveness of eyes in dim light
is a *tapetum*, a mirrorlike membrane behind the retina. Because
the retina is transparent, only a fraction of the light that reaches
it is absorbed and used, the rest passes straight through. The
tapetum reflects this unused light back so that the retinal receptors get a double dose. What is not absorbed by the receptors on
the return trip then passes out of the eye the same way it came
in, and this is what is seen by the motorist when an animal is
in the path of his headlights.

In any animal without tapeta, the light not absorbed by the
receptors is lost in the tissues behind the retina. The effect of the
tapetum, therefore, is like that of putting a mirror in contact
with the rear surface of a movie screen. None of the light that
passes into it is lost.

This sounds a simple device; but nature has in many instances
turned the tapetum into a highly sophisticated mechanical and
chemical instrument. In mammals, it consists of layers of cells
or fibers that are highly reflecting, and this is simple enough;
but in fish and crocodiles it is much more efficient in that it con-
tains crystals of *guanine*, a brilliant silver chemical such as is
seen in fish scales. In some bony fish and sharks, these crystals
work in a reciprocal arrangement with black pigment in such a

14. The guanine tapetum of a shark as it begins to appear when
the retina is peeled away from in front of it (top). It is fully revealed after the retina is removed (lower), leaving just a few small
folds on the head of the optic nerve where this enters the eye.

way that, as light increases in intensity, the pigment filters into the silver and obstructs its reflecting power, at the same time covering and shielding the sensitive rods. As the light diminishes, the pigment retreats and leaves the silver tapetum uncovered to amplify the light, and the rods withdraw from the pigment to receive the full benefit of this amplification. Even in some animals that do not have a guanine type of tapetum, a migration of shielding pigment and a movement of the rods take place.

Some sharks have an even neater arrangement. The silver crystals form plates at an angle of 45 degrees to the retinal cells, and the shielding pigment flows over these angled mirrors to cover them when the light is too bright, migrating back away from them when the light is reduced. There are many forms of tapetum throughout the animal world, and they vary in their efficiency, but they are all designed for the same purpose—to amplify the dimmest rays of light that reach the depths of the ocean or the dark recesses of the forests at night.

Because the tapetum increases the amount of light to the retina, the eyes of animals that are reflecting car headlights are being utterly blinded by glare, and the animal may remain transfixed because it is helpless. Motorists should not expect an animal necessarily to save itself. It is most likely unable to see where to go.

The kinds of tapeta found in mammals consist only of layers of cells or fibers behind the retina. What is so interesting in these two kinds is that the fibrous type is found only in the hunted animals, ungulates like cattle, deer, and goats; and the cellular type is found in hunting animals such as lions, cats, dogs, seals, and bears. Only animals active at night have tapeta, and in some mammals, at least, they are absent at birth and develop later.

Although there are so many animals with tapeta, there are also many without them, which nevertheless show eyeshine at

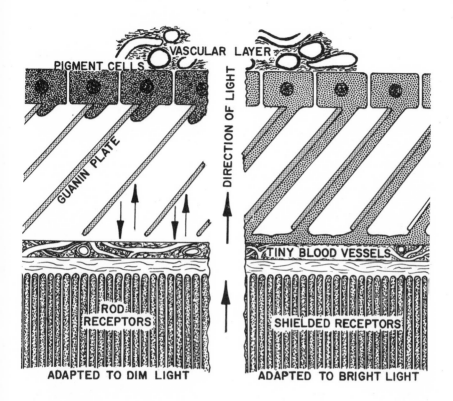

15. The guanine tapetum of the shark functions with an interesting and complex mechanism. It takes the form of silver plates behind the retina. In dim light, the light that the receptors do not absorb is reflected back by these plates so that the receptors get a double stimulation. In bright light, pigment migrates out from special cells to cover the guanine plates so that they do not reflect light. The pigment absorbs it. (Adapted from Franz.)

night if caught in car headlights. These are some of the rodents, some of the bats, some marsupials, snakes, toads, and birds. This eyeshine may merely be reflection by a fine membrane which covers the blood vessels behind the retina, and it is not known if it acts like a partial tapetum.

PROTECTING SENSITIVE EYES

All the advantages that come with exceptionally sensitive eyes would be useless if the owner ventured out into bright light without some form of protection. There are times when we like to protect even our relatively insensitive eyes with sunglasses. Preventing the eyes from being glared is most important because glare breaks down the chemicals in the rods to such a degree that these then take an excessive time to be resynthesized. The longer and greater the glaring, the longer the eyes take to recover, and this can be fatal in the wilds.

More than the protective pigment that passes over the tapetum is necessary to prevent glaring, and the first line of defense is in the contraction of the pupil of the eye as the light gets stronger. This is the aperture in the colored iris through which light passes to reach the retina. It is like the iris diaphragm of a camera, and the reduction of its diameter by a half reduces the amount of light passing through it to about one quarter. This contraction can be seen in our own eyes if we look into a mirror and then switch on a bright light, and we can calculate that if an 8-millimeter-wide pupil is reduced to 2 millimeters, the amount of light passing into the eye is then only 6 percent of what it would have been without contraction.

Pupil contraction in human eyes is very limited, and also very slow compared with that of many nocturnal animals. The owl's pupil response to light takes half the time ours does, both in contraction and dilation. An easy way to see a pupil work rapidly in a sensitive eye is to visit an aquarium where there is an

16. Most birds have circular and very active pupils, many of which are capable of extreme dilation and contraction. The dilated pupil of a bittern (top) contrasts markedly with that of the crowned crane (bottom), which contracted instantly as the bird walked out of shade into bright sunlight.

octopus, and take a bright flashlight. The animal's pupil response to the switching on and off of the flashlight directed at its eye will be very rapid and obvious.

Most pupils are circular, and this was probably the shape of the first pupils in the animal world. They certainly open wide at night, and some of them contract down almost to a pinhole in the glare of the day. But almost a pinhole is not always small enough. As fish developed more sensitive retinas for feeding at night, a circular pupil proved increasingly inadequate to protect these in bright daylight at the surface. They needed something that would contract to a smaller aperture than is possible with a bunching circular muscle formation. Greater contraction is achieved by a pupil the two sides of which will close like sliding doors. This has been accomplished by evolving a slit aperture, and a very large number of animals have this type, the slit being vertical, horizontal, or diagonal, according to their needs.

We see this slit-shaped pupil in almost all animals that are active in both bright and dim light. The sharks are particularly good examples, and on land the domestic cat is perhaps the most familiar. But there are countless others—amphibians, reptiles, and mammals. So far as we know, all pupils that contract to slits dilate to a circular or nearly circular shape in dim light. The cat shark's pupil shown in Figure 21d is almost unique. It may be even more effective than a conventional slit, the overlapping edges producing a minute pinhole at each end.

17. The barn owl (*Tyto alba*) and the Papuan frogmouth (*Podargus papuensis*) both have sensitive eyes protected by pupils that contract to very small apertures. But at night these pupils open wide to let in the maximum light. In fact they will open so much that the colored iris seems to disappear completely. If there is the slightest light or shadow visible, the owl will attack by sight, but in total darkness it will attack solely by sound direction.

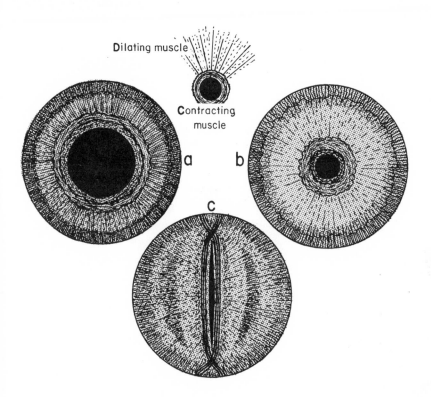

19. The iris of an eye has a circular contracting muscle that constricts or reduces the aperture of the pupil, and a radial dilating muscle that opens up the pupil. The constriction is limited because the contracting muscle bunches up around the pupil when it contracts. A slit pupil, on the other hand, frequently has two bands of contracting muscles, which close the pupil down like a pair of sliding doors, by straightening as they contract. The dilating muscles opposing these open the pupil to a circle when the two bands are relaxed.

18. The cuttlefish (*Sepia* spp.) is a cephalopod that hides in dark places and becomes active at night. The pupil of its sensitive eye is protected with a flaplike appendage which can, if necessary, close it completely but for the smallest pinhole during the daytime. The octopus (*Octopus cyaneus*), with similar habits, uses an active and highly contractable pupil to protect its eyes (*E*) in daylight. Inset is an enlarged eye, the pupil of which shows up white because light has been shone through from behind.

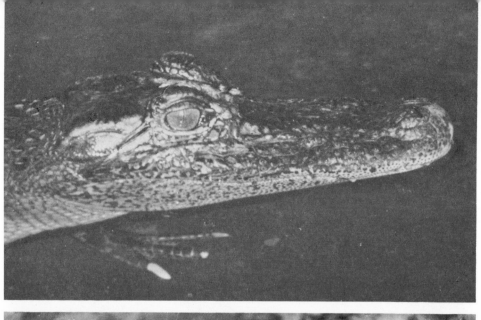

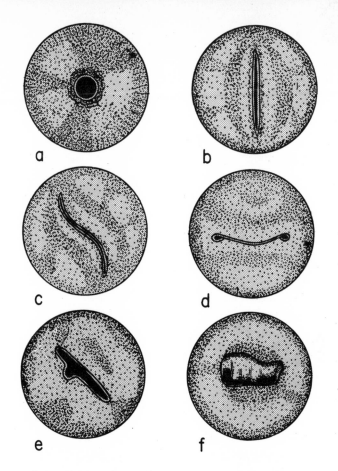

a   b

c   d

e   f

21. Some of the shapes of pupils used to protect sensitive eyes. (*a*) A standard circular pupil about as small as it will close relative to the size of the iris. (*b*) That of the dog shark (*Mustelus canis*) and (*c*) that of the carpet shark (*Orectolobus maculatus*) will virtually close completely. (*d*) The pupil of the cat shark (*Scylliorhinus* spp.) will leave a tiny pinhole at each end when it contracts, and (*e*) that of the Port Jackson shark (*Heterodontus portusjacksoni*) will leave one small aperture where the notch is in the edge of the pupil. (*f*) The pupil of the fiddler ray (*Trigonorrhina fasciata*) will almost completely close as the curtainlike upper edge of the iris drops down further.

20. The crocodile (*Crocodilus porosus*) and the alligator (*Alligator mississippiensis*) are very active at night, and their eyes are unusually sensitive, so they are protected in the light by fine slit pupils. The freshwater tortoise (*Emydura macquari*), on the other hand, is sluggish, and not a hunter. It does not leave the water much during the daytime, so protection is not so essential for its eyes, and it retains the less active circular pupil.

Some of the ungulates, of which the horse is a good example, have a shading membrane suspended from the top of the pupil. It is really an extension of the upper part of the iris, which, as the pupil closes, meets the lower edge of the pupil, leaving an un-obscured area of pupil at both the front and rear, where a hunted animal needs its greatest consciousness of what is happening around it. Figure 22 shows this kind of membrane in a relaxed state of semidilation, and it will be seen that when contracted the pupil greatly resembles that of the cat shark.

FITTING LARGE EYES INTO SMALL HEADS

The advantages of large eyes at night have always been exploited by the animal world when possible. They produce a larger image, and this therefore covers more receptors, giving it a much finer quality and definition. But a large image also spreads the light over a greater area, making it less bright. This would create no problem in the daytime, but it does at night, and is overcome by using pupils that dilate more widely and allow more available light to enter the eye. Increasing the diameter of the pupil four times increases the amount of light entering the eye sixteen times.

In Figure 23, the sizes of images in different-sized eyes are illustrated with two conventionally shaped eyes, spherical ones; but small heads limit the size of eyes they will hold, so another way of producing large images has been found by many animals.

22. This aoudad (*Ammotragus lervia*), like all wild goats and sheep, has a sensitive retina which is protected by a pupil that will contract down to a horizontal slit. The pony (lower left) has a horizontal pupil too, but this is aided by a central curtainlike projection called an umbraculum. (Lower right): Here an eye has been opened so that the rear surface of the iris and the umbraculum are more easily seen.

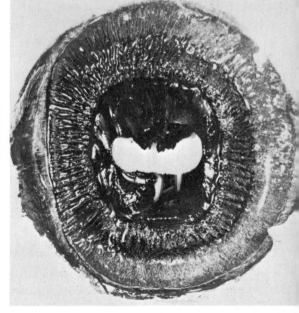

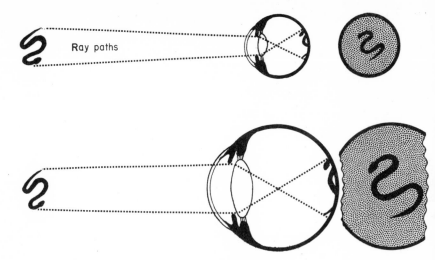

Ray paths

23. Larger eyes produce larger images, and if the receptors are approximately the same size in both, then a larger eye also produces a clearer image. Frequently the large eye will also admit more light, which is very important at night. One of these eyes is just twice the diameter of the other, but the image in it is more than twice the size of the other image. The pupil of the larger eye has a diameter three times that of the smaller, but it admits nine times as much light. So a large eye has many advantages at night.

Eyes are developed as large as the head can possibly contain, and then the back surfaces are shaped as though they were segments of even larger eyes. This usually reduces the extent that eyes can be moved, and sometimes they become completely immobile, but if an animal has a very mobile neck, or needs no wide field of vision, it loses little by not being able to move its eyes very much. Large eyes also mean, however, that the animal has very little room in its skull for a brain, so that this is strictly limited in size.

Although the eye shapes shown in Figure 24 are so different, their effectiveness is somewhat similar, partly because the curves of their backs are similar. The great length of the owl's eye (front to back) produces a large image because its lens and cornea are a long way forward from the retina. In the fish, the whole system is different because the eye is in contact with water which, having a greater refracting power than air, reduces the power of the cornea where it touches this. The image obtained is large, however, because the lens inside the eye has a greater power to compensate for the power lost by the cornea, and because all visual images are larger in water than in air.

To illustrate this simply, perhaps it can be said that if in the owl, for instance, the magnifying power of the cornea is 5, and that of the lens is 3, the resulting image has a value of 15, because

24. An eye that produces a large image will see better in dim light than one that does not. A large eye is needed for a large image, but nature has overcome this problem in small heads in several ways. (*a*) is the cut-away plan of an owl's head showing eyes which, though not appearing particularly large at the front, have a back surface curve that is equivalent to an eye the size of the dotted circle. Such eyes cannot be moved in their sockets, so the owl has a very flexible neck, and can turn its head through 360°. Smaller and less nocturnal birds are more like (*b*). (*c*) shows that a fish has a similar way of producing a large image. (*d*) is a telescopic eye found in some deep-sea fish. It gives great magnification, but a very limited field of view. The images in all these eyes are equivalent in quality to the images that would be produced by spherical eyes (like ours) the sizes of the dotted circles.

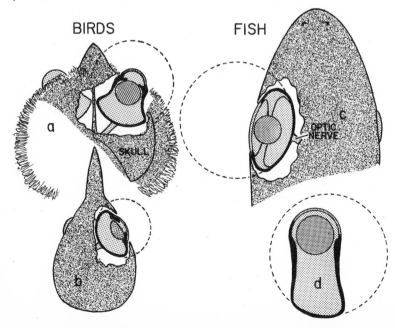

BIRDS   FISH

the lens further magnifies what is produced by the cornea. If in the fish the magnifying power of the cornea is nil because it is in contact with water, but that of the lens is 15, the resulting image value is the same.

Many nocturnal mammals have unusually large eyes, but they also have skulls capable of accommodating them. Figure 34 of the Philippine tarsier is an example. Another, similar animal is the douroucouli (*Aotes trivirgatus*) of Central and South America, which is probably the only truly nocturnal monkey.

WHAT HAPPENS IN THE BRAIN

Because there are so many factors involved, what happens to a visual picture by the time it reaches the brain of each kind of animal cannot be described here, but we can take a very simple example, much simpler than that of our own eyes, to show the path taken by the impulses from each eye, and illustrate this with Figure 25.

Let us assume that the two eyes shown are those of a snake on the hunt at night. It sees a small gecko silhouetted on a rock, and, as a result, an image of this gecko is formed in each eye. These images, which are in reverse, are transmitted along optic nerves to opposite sides of the brain, each one arriving at a center where some complicated integration takes place before it is relayed on to a part of the cortex, situated toward the back of the brain.

Here, the two images are again reversed and combined into one single image the right way around and up. Though this is the simplest pattern of coordination in the higher animals, it may still seem unnecessarily complicated, but it is not. To register such diverse things as shape, density, movement direction, and a host of other subtle factors involved in vision, a very sophisticated computer system is required, and all this has to be included in the

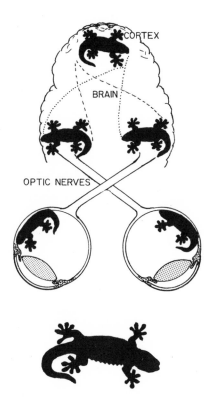

25. The images formed in the eyes are always reversed. In most animals the images are transmitted to the opposite sides of the brain, where they are partly decoded and relayed on to the cortex, again being reversed on the way. In this cortical area, both images are combined into a single image that is more than the sum of the two originals, because it has a stereoscopic form produced by the two different angles at which the eyes see the object looked at. The combination of two images into one in this way also doubles the intensity or brightness, a very useful factor at night.

short pathways which, in many animals, may be less than an inch in length.

The combination of the two images in the cortex accomplishes at least two aims: it produces a stereoscopic effect, giving what is seen the qualities of depth, thickness, and solidity, and it also increases the intensity of the image. We can appreciate this if we look at something dim and alternately close and open one of our eyes. When only one eye is open, the object is not nearly so bright as when both are open. When both are open, double the amount of stimulus is reaching the image in the cortex.

Our own brain patterns and those of the highest animals are of course much more complicated than the one just described; not only do an animal's senses adapt themselves to new ways of life, but its brain has to develop along new paths at the same time to make use of these new adaptations.

Lights in the darkness

Some eyes have to function in almost perpetual darkness—in caves, and particularly in the ocean depths where little or no light penetrates at any time. Fish that live in these environments are faced with an evolutionary choice: to develop eyes that are remarkably sensitive, or to discard them altogether and use some other sense to detect food and mates. Perhaps most develop large and exceedingly sensitive eyes with which they can pick up the faintest trace of light reflected from another moving creature or from the luminescent organs this creature may have. But there are many fish that get along quite well without eyes at all. Others develop tubular-shaped eyes which, like large spherical ones, provide a very large image of what is seen, and at the same time accept the faintest stray light.

Luminous Organs: Recognition Displays; Finding a Mate

Deep in the oceans, where only the faintest rays of light are ever present, all creatures are likely to prey on one another, and it becomes essential for them to recognize their own species and also to differentiate between the sexes. In well-lighted waters, where fish can see each other clearly, this recognition is through shape, color, or markings; but down in the depths it is doubtful if even a silhouetted moving shape could be reliably identified by the sharpest-eyed fish. So patterns of tiny lights distributed over

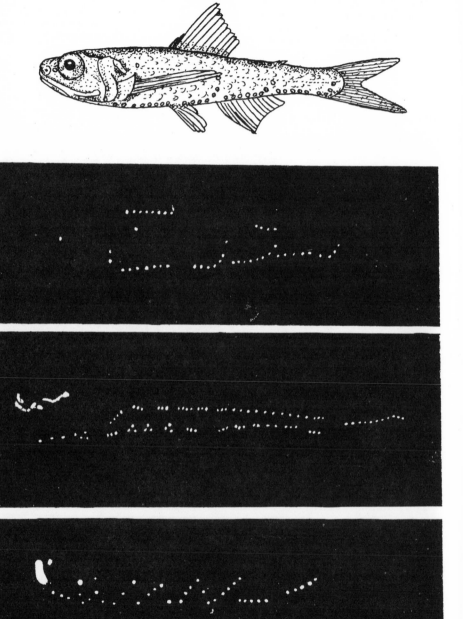

the body, differing in the sexes, have become widely adopted by countless deep-sea fish. These lights are known as *photophores*, and we find that at some ocean levels up to 95 percent of the fish use such light organs in one form or another, either flashing them on and off at intervals or keeping them continuously lighted.

The ocean is divided into definite depth zones. At a depth of 200 meters (between 600 and 700 feet), what is known as the *pelagic* (upper or coastal) zone merges into the *bathypelagic* zone (*bathy* is from the Greek word meaning "deep"). This zone ends at about 2,000 meters (between 6,000 and 7,000 feet). Between this and what is known as the *abyss* (deepest water) is a kind of "no-man's-water" where life is far from plentiful, but it does exist to some extent.

Creatures living in this zone have some unusual problems when they need to find food or to find each other for mating, but such an invariable and well-organized system of recognition light displays has been developed by them that fish can be classified into species by the patterns of their luminous organs. These never vary within a species, or one sex of a species.

The light patterns, and the flashing intervals used in many of them, keep fish together in groups, bring the sexes together and, less intentionally, signal the owner's presence to predators. However, there are also light systems designed to fool predators. And, in turn, predators use luminous lures to bring victims within their reach. Luminous organs that are not displayed con-

26. A lantern fish (*Diaphus danae*), and immediately below it the pattern of its light organs as seen in the dark. Beneath this, the center set of lights are from another fish (*Aristomias grimaldii*), in which many of the lights are in sets of three, and in which a very bright, unusual pattern is seen on the head and beneath the eye. The bottom set of photophores is interesting for the bright headlight just in front of unusually large eyes.

stantly are usually flashed on momentarily in response to an awareness by sound or other sensation that another fish is nearby. A responding flash of recognition from the other fish which shows the same species pattern may then stimulate a full display of lights or repeated flashing in both of them.

Some fish flash their lights at precise intervals—on for so many seconds, then off for so many, a recognition or signaling pattern which will be constant throughout the species and quite different from all other species, even closely related ones. The flashing or lighting of these organs can be stimulated artificially by adding ammonia to the water or by discharging an electric current into it.

Because fish living at great depths during the daytime often rise to upper levels at night, we are sometimes able to see or even catch small specimens studded with light organs. Some of the coastal waters of Australia abound with them, as does the Pacific Coast of North America, but usually they come really close to the surface only on moonless nights. They are seldom more than twelve inches long, and their daily upward and downward migration is evidently in pursuit of food. They follow the rhythmic migratory movements of small invertebrates or plankton, which form fluctuating layers in the sea. Just what produces the urge or need for such forms of microscopic life to change their level each day is not known. It can hardly be temperature change, because temperature in the depths is constant. Nor can it be a response to light and dark cycles, because they always re-

27. The hatchet fish (*Argyropelecus* spp.) has a thin body like an ax blade, and in spite of its mere two-inch length it carries quite a battery of lights. The pattern they show in the dark is below. As this fish's body is quite transparent, it must also glow to some extent from diffusion of the light, but the predominant direction of the beams from these light organs is downward, while the eyes point upward, suggesting that recognition of their own species is from below.

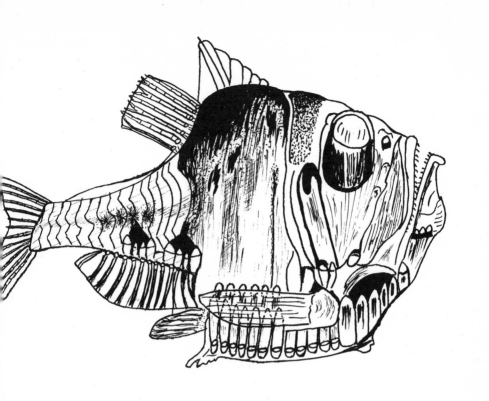
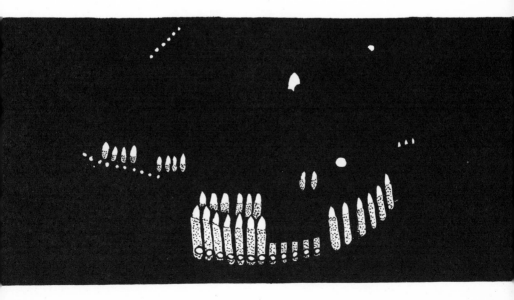

FORMATION OF LIGHT

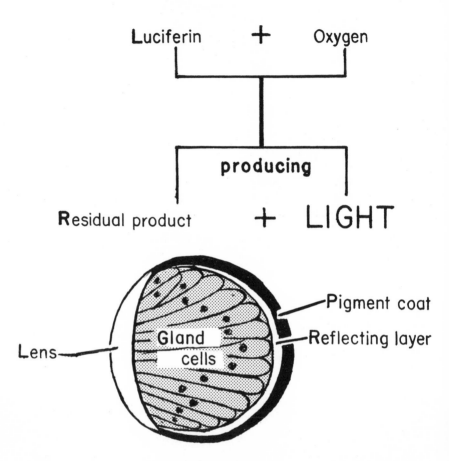

Luciferin   +   Oxygen

producing

Residual product   +   LIGHT

Lens

Gland cells

Pigment coat

Reflecting layer

28. It was shown in Figure 10 how light affects the chemical in the retinal receptors of the eye, breaking it down to a protein and creating an electrical response, the protein then combining with vitamin A to start the cycle over again. Photophores have a somewhat similar action. In this case, oxygen and not vitamin A sets up the chain reaction that ends in light and a residue. It combines with a chemical called luciferin, and the light that results is the most efficient known. Some of the photophores in which the action is produced are very similar to eyes, which are just as effective for passing light out as for receiving it.

main in darkness. It may be a nonrhythmic biological timing mechanism.

Not the least interesting feature of light organs is their extreme efficiency, all the energy produced being used as cold light. The different ways the light is produced and the ways in which it is controlled so that it shows constantly, flashes, or appears in a burst are quite remarkable. The most highly developed kind of luminous organ is something like a combined eye and gland of minute proportions; like a gland because it has cells that secrete a fluid, and like an eye because it has a focusing lens that produces an almost parallel beam, an iris to control the amount of light, a reflecting layer at the back, often of silver guanine, and a layer of pigment enclosing the whole organ to prevent light from radiating into the body.

The gland secretes a compound called *luciferin.* This is a phosphorus compound, and when it is broken down by oxygen in the presence of an enzyme called *luciferase,* it glows. The required oxygen is obtained from a plentiful supply of blood vessels serving the gland. This is a neat arrangement, for the blood vessels can be constricted by tiny muscles around them to shut off the blood supply. This in turn stops oxygen from reaching the gland and so extinguishes the light.

Only recently has man been able to duplicate this manufacture of chemical light artificially. Chemical light requires no electrical connections or batteries, and its advantage is that it can be used in liquid, powder, pellet, flake, and sheet form, all activated to become luminescent when they come in contact with air or oxygen. It does not burn and produces no heat; and it can be turned off merely by closing the container it is in to cut off the oxygen in the air. Once again man has been able only to copy what nature has long been doing.

Another and simpler kind of light organ found in fish consists of groups of gland cells derived from those that secrete the skin slime. These are sometimes grouped in depressions, and as the secretion of skin slime leaves the gland cells and comes in contact with the water it takes up oxygen and glows. A third form is made up of colonies of phosphorescent bacteria living in pits in the skin, and these glow when oxygen reaches them either from blood vessels, which can be controlled by muscles, or from the seawater.

We would not expect the lights produced by these organs to be colored, because different colors require chemical variations or different structures, possibly with filters. But there are nevertheless variations in color. Most are blue or blue-green, but there are also a few species of fish that use reddish, pale yellow, yellow-green, orange-purple, or blue-white light. The most interesting feature of these color variations is that, in those fish that have been thoroughly investigated, the wavelength of the light from the light organs in a particular species coincides exactly with the wavelength that is most readily absorbed by the sensitive receptors in the eyes of that species. This means that the eyes of any species are most sensitive to the hue of light from the light organs of its own kind, and so strongly supports the theory that light organs are to enable fish to recognize their own kind in the dark, and thus to remain in groups or to find mates.

There are many species of deep-sea fish that have luminous organs around, or at some point adjacent to, each eye. Sometimes these organs are very large, and their purpose must be more specialized than that of most other light organs because they can usually be obscured, either by rotating them with muscles that draw them down into pigmented pockets or by the drawing of a dark, curtainlike membrane across them, just as eyes can be closed with eyelids. These organs usually have good focusing structures, but sometimes even the simple pits of luminous bacteria are similarly controlled by shutterlike mechanisms. This has the same effect as the constriction of the blood vessels to cut off the oxygen to the bacteria or the gland cells, except that, when light organs are obscured by blinds, they probably shine continuously even under cover, whereas, when controlled by the blood vessels, the light is not continuous.

Perhaps related to these light organs near the eyes are other light patches in some fish which, like the colored eyespots near the tails of certain surface-living fish, are probably designed to confuse predators. Sometimes they perform a dual purpose, being used also to flash signals, perhaps as the firefly does in the mating season. In some species, the males flash but the females do not, and it does seem that the males, as with some birds, try to attract the attention of predators away from the females, which are thereby left in relative peace to produce the young that will perpetuate the species.

Most remarkable of all are glandular light organs on the sides of the head which can expel luminous clouds into the water, thus confusing an enemy while the fish escapes. This is rather like the way the octopus or squid ejects an inky blob into the water and escapes while an enemy is investigating it. Even invertebrates have managed to evolve this habit. A shrimp (*Acanthephyra purpurea*), which lives in the lightless depths, releases a cloud of luminous fluid to blind or confuse an enemy while it escapes.

Angler fish with baits dangling on a flexible first dorsal ray or from barbels are well known to fishermen. These fish are by no means confined to the lighted parts of the oceans. There are great numbers of them in the depths, too, and they carry illuminated baits. These may be yellow, yellow-green, blue-green, or orange, and some can be flashed on and off by constriction of the blood vessels in them, just like other light organs. Usually they appear to consist of phosphorescent bacteria, which glow in response to oxygen from the blood vessels. One angler fish even has a luminous bait inside the mouth and others have glowing teeth, probably caused by bacteria in a mucous covering. There is still much to be learned about this.

The luminous patterns adopted by fish can be numbered in thousands, as many as there are species of fish using light. And sometimes there seem to be thousands of these tiny lights even on the smallest fish. A double row down each side, like portholes, is a very common pattern, sometimes below the lateral line or even on the fish's underside. These lights may be evenly spaced, or in groups of straight threes or any other number, or in triangular groups. They may be on the fins, too, or, as in the viperfish (*Chauliodus* spp.), even inside the mouth. Some of the little lantern fish that come near the surface at night have definite headlights, and some also have powerful flashable lights close to the tail.

That luminous organs are not a recent development in fish is obvious when some of the deep-sea sharks are examined. These are very ancient types, and apart from the fact that they do not appear to have reflectors, their light organs are similar to those already described, glandular in form, and with lenses and pigmented backs. Sometimes the light organs are unusually small and so numerous, even on the pectoral fins, that they must appear more like a glowing outline to other fish than as actual rows of lights. They are less specialized than those of the teleost fishes, but because sharks have such sensitive eyes, they must be adequate for species recognition.

Luminescence in the sea is not confined to fish and crustaceans. In many parts of the oceans, anything moving through the water is seen at night because countless minute organisms light up when disturbed. Frequently these even light up areas of water enough to make fish visible. There are also corals and plants that fluoresce in various colors, and there must be some purpose in this, although we do not yet know of any living creatures that have eyes specialized to respond to this fluorescence.

On land there are lichens and mosses that fluoresce under certain circumstances, and controlled light displays like those described on fish are seen in fireflies and glowworms. As among fish, these displays are for recognition in the dark, and to bring together the opposite sexes for mating.

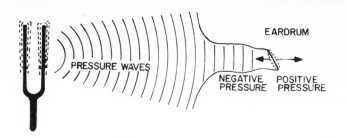

29. Sound vibrations produce pressure waves in the air. When the prongs of a tuning fork vibrate toward the eardrum, they create a pressure on the drum that is higher than the atmospheric pressure, and when they rebound in the opposite direction, the pressure on the drum is less than the atmospheric pressure. The responding vibrations of the drum are therefore said to produce positive pressure when they are inward, and negative pressure when they are outward.

30. Sound is collected by the pinna of the ear, and directed into the outer ear canal. It strikes the eardrum, which vibrates and passes the vibrations to the three bones in the middle ear, which in turn connect with the inner ear. In the small box (upper right), this part of the chain is enlarged. D is the eardrum; M is the malleus; I is the incus, and S the stapes. The stapes is immediately over a window into the inner ear, where the nerves to the brain are stimulated. Reptiles have only a stapes bone attached to the eardrum (lower right), but in either case the ratio between the area of eardrum and that of the foot of the stapes controls the amount of amplification the system provides.

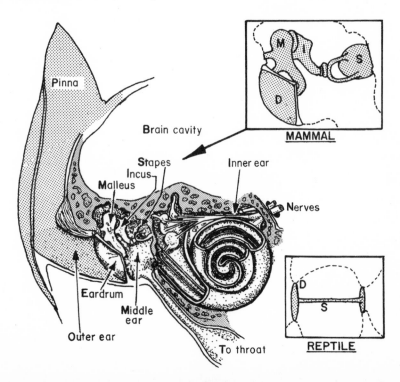

Ears and superears
for the dark

There are more sounds in nature than we can hear. In fact, of all those used by the animal world, we may appreciate less than 10 percent. The rest are pitched higher than our ears are able to record. Though the human hearing range is said to be from a frequency of 16 cycles (vibrations) per second (cps.) to 30,000 cps., it seems certain that 23,000 cps. is the upper limit for most children, and 20,000 cps. for most adults. Probably few adults can hear any sound with a frequency higher than 10,000 cps., or lower than 50 cps.

The middle C on a piano has a frequency of 256 cps., and the highest tone achieved by a musical instrument is about 4,800 cps., so naturally we would think a tone of 10,000 cps. very high indeed. And yet the bat can easily hear sounds as high as 100,000 cps. The top limit of its range is almost certainly in the region of 300,000 cps.

Scientists in Russia have claimed that it is possible for the human ear to recognize ultrasounds, but the conditions of their experiments to prove this were not those of natural life. The sound source was placed on the forehead and behind the ears of each person tested, and this means that the vibrations were conducted through the skull and not by the customary route through air and the outer ear. So, even though the brain was proved capable of registering high frequencies, it would not be able to do so under normal conditions. It was claimed that the subjects used, all forty years of age or less, could distinguish sounds up to 225,000 cps., 10 times higher than by the normal route. Underwater swimmers might conceivably have these high frequencies conducted to their recording centers through the skull's contact with the water if such frequencies could be propagated in water with sufficient volume.

The human ear can withstand a great range of volume or loudness; from a point where a sound can only just be heard (the threshold) to an amplification of ten billion times. Intense volumes such as this are fortunately not encountered normally; they can produce permanent damage to the delicate ear mechanism if they are prolonged. We have less information on the volume of noise or sound that other animals can stand. However, though this must vary greatly, some of the animals on which observations have been made do not seem bothered by a degree

31. The fact that so many animals eat termites indicates that there is nourishment in these insects, but it also emphasizes the acute hearing of those animals that hunt them. These three mammals are all termite hunters; the bat-eared fox (*Otocyon megalotis*) at the left, and the aardwolf (*Proteles cristatus*) at the right hunt by listening for the movements of the termites in their burrows. The aardvark (*Orycteropus afer*) in the center both listens and smells. In all three animals the ear pinnas are large for picking up the slightest sounds.

32. (Top): This animal is typical of small placentals and marsupials, which rely on their hearing for much of the information they need on what is happening around them in the dark. The aye-aye (*Daubentonia madagascariensis*) has very naked and dishlike ear pinnas capable of being highly directional, even to the extent of detecting insects in wood. If the pinnas were lined with fur, sounds would be dampened, so there is purpose in their nakedness. (Below): The senses of the Virginia opossum (*Didelphis*) are balanced more evenly for its night life. A prominent snout indicates a well-developed sense of smell.

33. Those nocturnal animals with small and often not very useful eyes invariably make up for this with unusually sensitive ears. This is obvious by the size of the pinna, the erect trumpetlike outer ear, which, like a radar dish antenna, can move in different directions and reflect the slightest sound down to the eardrum within. The long-eared bandicoot (*Thylacomys lagotis*) and the hopping mouse (*Notomys mitchelli*) are good examples. Not unlike this hopping mouse are the dwarf kangaroo rats of the western United States.

34. Philippine tarsiers (*Tarsius syrichta*) are exceptionally well equipped for night life; they have two senses highly developed. Their large eyes are sensitive and have widely dilating pupils; and their ears are particularly sensitive. These animals can leap through the trees with great speed and precision. (Lower): The Euphrates jerboa (*Allactaga euphratica*) has much the same characteristics, large sensitive eyes and ears. Though a ground dweller, it leaps like a tarsier, both animals being able to cover up to six feet at a bound, in spite of their small size. *Photos by permission of the Zoological Society, London.*

of loudness that would be uncomfortable for us. For instance, some of the underwater calls made by seals can be very uncomfortable for skin divers, but they evidently are not for the seals.

How Animals Receive and Use Sound

It is natural that animals that are active in darkness rely greatly on their hearing and also need to respond to a wider range of sounds than we can recognize. So it is understandable that the vast majority of mammals have larger ear pinnas than we have. The *pinna* is the trumpetlike outer ear, and the word is from a Greek word used for some seashells. The recognition in some mammals of sounds pitched up to hundreds of thousands of cycles per second is in great contrast to the responses of birds, most of which, being active during the day only, rely mainly on their vision, and so can probably not hear sounds above 10,000 cps. The few nocturnal birds are exceptional to this.

A large ear alone does not guarantee an appreciation of a wide range of sounds or volume. It only serves to pick these up and direct them into the middle and inner ear chambers where the auditory mechanisms are. Just as important are highly developed acoustic or hearing centers in the brain, and an animal that relies

35. The douroucouli (*Aotes trivirgatus*) of Central and South America has similar eyes and habits to the tarsier. It is probably the only truly nocturnal monkey.

36. Three silent nocturnal fliers. The European barn owl (*Strix flammea*) has excellent night vision and hearing; the gliding possum (*Schoinobates volans*) has good night vision, and its eyes show considerable shine if caught in the light of a flashlight. It can glide long distances in the dark. The American counterpart of this animal, although not an opossum, is perhaps the flying squirrel (*Glaucomys*), which is the only nocturnal squirrel known.

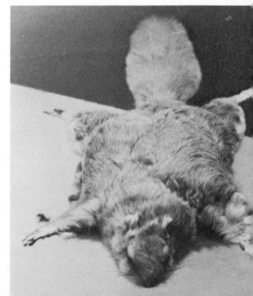

greatly on its ears will also have a much greater area of its brain devoted to integrating and understanding what it hears. This applies even more to those animals that use *echolocation* or *sonar*, in which sounds are transmitted by the animal and the reflected echoes are interpreted so accurately that even in complete darkness the smallest creatures can be detected and captured.

Some knowledge of the function of the ear helps in understanding the complexity of interpreting sounds. The ear is divided into outer, middle, and inner chambers. Sound vibrations are collected by the outer ear, which acts somewhat like a dish antenna, and directed on to the taut eardrum, the pressure they create causing this drum to vibrate also. This resulting vibration of the eardrum is so slight that it can be measured only in millionths of an inch, but it is enough to activate three small bones in the middle ear behind it.

The three bones are known as the *malleus* (hammer), which is in contact with the back surface of the eardrum, the *incus* (anvil), and the *stapes* (stirrup), shaped like a rider's stirrup. The footplate of the stapes covers a window into the inner ear, so the bones conduct the eardrum vibrations right to the inner ear. This window has an area of only 3.5 percent that of the eardrum, so the system amplifies sound about 28 times.

The inner ear is a very complex arrangement of fluid chambers and canals in which the sensitive endings, or receptors, of the auditory nerves to the brain are located, and part of which controls the balance of the entire body. When the vibrations reach these nerve receptors, they stimulate them to form electrical impulses, just as the receptors in the retina of the eye form electrical impulses in response to light vibrations. The tiny currents created pass to the brain centers, where they are interpreted as sound.

JUDGING DIRECTION

The mixture of sounds must be decoded and analyzed, too; background or unwanted noises must be filtered out; direction must be recognized by rapid measurement of the difference in sound volume reaching the two ears; adjustments must be made for loudness, and experience and memory must be involved to recognize and label every sound. It is not even as simple as this, but it is all we need to know to understand the basic principles of hearing.

The pinna is almost always large in night-foraging animals. It gathers sound like the mouth of a trumpet, and directs it into the passage to the eardrum; thus the larger it is, the more sound it collects. And its shape is important too, for certain sounds have their pressure effect changed by being directed into the ear canal at particular angles. Perhaps the sounds which are most important to an animal can be either amplified or separated from others better in this way.

Placing a large seashell to our ear gives us the impression that we can hear the sea in it. This is only because the shell collects many sounds that we do not normally hear, and adds them to the sounds we do hear ordinarily but manage to ignore. Knowing this, it is understandable that experiments on animals in which the pinna of the ear was removed and the brain responses to sound were measured electronically showed that the level of hearing was greatly reduced, and judgment of direction was interfered with.

Echolocation by bats

The hearing of insect-eating bats has had considerable attention from research workers for a long time because these mammals make use of a system of echolocation that enables them to locate and catch tiny insects while in flight in complete darkness. More recently the marine mammals—dolphins, whales, seals, and sea lions—have been given much more study because some of them are exceedingly sophisticated in their use of natural sonar (echolocation in water). An unexpected discovery that some nocturnal birds use a similar system has made scientists realize that they have a great deal to do yet before they will uncover the full extent of the use of echolocation. It is becoming obvious, too, that there are even fish that make use of it.

The advantage of echolocation is that it permits speedy movement in the dark; as fast, in fact, as would be possible in the light with perfect vision. Now, with modern electronic technology, it has become possible to make use of our knowledge of this by designing miniaturized instruments for blind people, which will enable them to get around by using echolocation instead of vision.

Bats appear to have changed very little in their more recent period of evolution, and this shows that their way of life is well suited to survival. At night, insects are plentiful and predatory birds few; so, apart from disease, those bats that do not rest in

37. The large ears of these two bats, the false vampire bat (*Macroderma gigas*) and the Australian long-eared bat (*Nyctophilus geoffroyi*) will pick up echoes from mosquitoes flying in the open, which reflect their sound transmissions. The false vampire bat shows a flap above the nose that directs the sounds outward in a concentrated beam. See Figure 42.

trees during the day, where they can be found by exploring snakes, have few enemies to disturb a well-ordered life.

Not all bats use echolocation. Most fruit bats and the flying foxes have little need for it. Their night vision is by no means useless, and during the day most of them hang in the dense forests, flying off in vast formations to their feeding grounds in the early evening. But the insect-eating bats all use echolocation in various degrees, and some of them are so sensitive that they can "home in" on insects as small as fruit flies, catching them at a fantastic rate. So fast are bats' movements that high speed cinematography has shown one catching two flies in half a second; another record from laboratory observations revealed that one caught 220 flies in fifteen minutes.

How It Works

Just how is this done? When cruising, the bat sends out pulses about 10 times a second, and as soon as it receives back an echo suggesting food, the pulses are speeded up as the object is approached, until there are about 200 a second. This increasing speed appears to improve the accuracy of the approach. The number of pulses transmitted each second is not to be confused with the frequency of the sound in cycles per second. The frequency, which is also variable, is really the pitch of the sound. For instance, while cruising and giving off the slow pulse rate (10 a second), the sound may be pitched at 100,000 cps.; but when an insect is detected, this drops to 40,000 cps.; and when near the target, it slides down through 30,000 to 20,000 cps. So it seems that as the number of pulses to the second increases, the frequency of vibration drops, and this scaling down of frequencies as an object is approached is evidently a device used by the bat to be certain it is listening to its own echoes, and not those of another bat.

It is therefore related to efficiency; it is not because frequency of vibration and pulse rate depend on each other within the range used by the bat at the speed at which it flies, though this would be so at some frequencies. However many pulses are

38. The nocturnal habits of these two bats are quite different, and this is immediately obvious. (Above): The giant fruit bat, or gray-headed flying fox (*Pteropus poliocephalus*) has small ears, a prominent snout, and large eyes, which make it by no means "as blind as a bat." These bats fly in steady formations of huge numbers to the feeding grounds, where they clamber in the trees and consume large quantities of fruit. The long-eared bat (*Plecotus auritus*) has small and less useful eyes, but immense ear pinnas for picking up the slighest echo from its own echolocation system. It catches in flight and feeds on insects as small as mosquitoes.

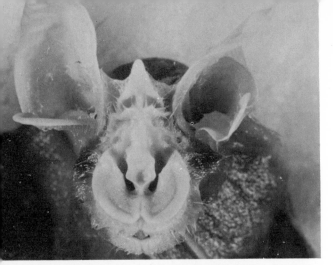

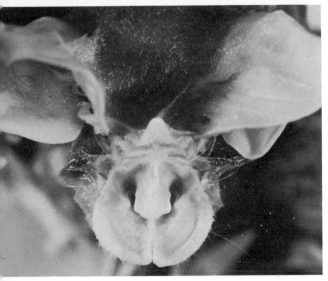

39. The head of the horseshoe bat (*Rhinolophus megaphyllus*) from the front and above, showing the way in which the nostrils are located within the horseshoe so that sounds can be concentrated forward as the bat is flying. Similar to the horseshoe bat is the American lumpnosed bat (*Corynorhinus*), which ranges from Virginia to the Pacific coast.

emitted each second, the frequency *can* remain the same, and two simple calculations may help in understanding this:

1. If a note of 10,000 cps. is being transmitted, and it is released in 100 pulses each second, with an equal space between the pulses, then each pulse will be 1/200 of a second in length (100 pulses and 100 spaces in a second).

   So, as $\dfrac{10,000}{200} = 50$, each pulse will contain 50 vibrations of that note.

2. Similarly, if 10,000 cps. are emitted in 500 pulses each second, with an equal space between them, the pulses will be 1/1,000 second long, and as $\frac{10,000}{1,000} = 10$, each pulse will contain only 10 cycles or vibrations.

But the note will still be 10,000 cps.

JUDGING DIRECTION

A bat's judgment of direction is quite uncanny. It is achieved in the same way as with us or any other mammal, but it is much more sensitive and precise. The sounds entering the two ears are compared, and their amplifications analyzed in the hearing centers of the brain in thousandths of a second. If one ear is plugged, the bat can avoid only large obstacles, and if the plug is permanent, or there is permanent injury, the bat will starve to death, unable to locate insects accurately enough to catch them.

40. A section through the head of a horseshoe bat showing large chambers, or sacs, in the respiratory system. These are evidently involved in the amplification of sound.

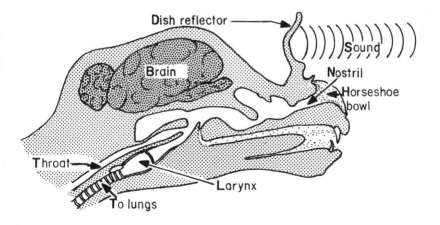

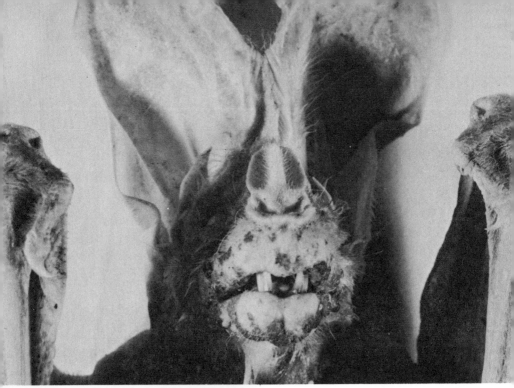

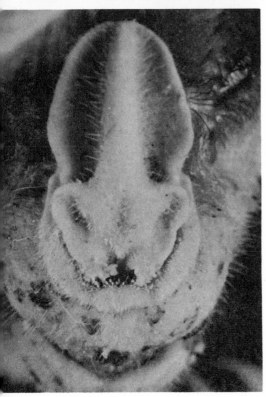

41. The false vampire bat (*Macroderma gigas*) has a nasal appendage which functions somewhat like that of the horseshoe bat. It is shown enlarged at left. In the lower front part of the ear pinna (top picture) a pointed appendage known as a tragus can be seen. This aids in directing sounds down the ear canal by reflecting them as they bounce from the large pinna.

If the mouth is forcibly closed, the echolocation system breaks down. Any bat with damage to its larynx would not only starve, it would soon kill itself by crashing into obstacles. If a bat that uses echolocation is born with a defect of any kind in its ears or hearing centers in the brain, it will never survive.

There is no doubt that many insects have a system sufficiently similar to that of the bat that they can take some measure of evasion with it. The moth's hearing, or whatever the system is with which it responds to sound vibrations, seems to be as efficient as that of the bat.

Any bat that has large ears, and this will comprise most of them, is certainly likely to use echolocation for finding and catching prey and flying in the dark. Some outstanding examples of large-ear development are seen in the pallid bat (*Antrozous pallidus*) of the southwestern United States, the spotted bat (*Euderma maculata*), the lump-nosed bats (*Corynorhinus*) which occur over most of the United States and in British Columbia, and the large-eared bats (*Macrotus*) of California.

JUDGING DISTANCE

The insect-eating bat's judgment of the distance to an insect or object is controlled by the interval between sending out the pulses and the reception of the echoes. As this is usually no more than a thousandth of a second, it may be another reason why the frequencies decrease as the animal approaches its prey. The bat appears to catch prey by scooping it into its wing or tail membrane, and then taking it into the mouth from there. It never takes prey straight into the mouth, but, according to the photographic records made so far, the whole operation is as swift as if it did.

Such bats are able to use their echolocation even in a great volume of interference, and laboratory experiments have shown that even very high levels of man-made noise will not disturb their performance any more than the sounds from a surrounding flock of other bats. Efforts have been made to catch bats in nets by confusing them with man-made noise, but it seldom works.

If we could hear the sounds made by a bat, most of which are above our own level of appreciation, we would find them very loud, by no means the tiny squeaks expected from so small a creature, and obviously this is necessary to get echoes from such small insects as mosquitoes and fruit flies. Some bats even have a tiny muscle in the vicinity of the stapes bone that increases the pulse pressures reaching the ear when these are below 50 a second, and this further amplifies echoes from more distant targets to a considerable extent.

Although some bats' brains weigh no more than thirty-five thousandths of an ounce, they contain relatively large and highly evolved hearing centers, far more advanced than our own. There are species differences, however, just as there are species differences in their voices, which vary in different localities. Bats do not always cease to be vocal when they stop flying. Many bats emit ultrasounds so long as they are alert. They test their surroundings with their ears, just as we do with our eyes.

SOME DIFFERENT BATS

Because of strangely shaped membranes on the nose, the two groups of bats known as horseshoe bats and false vampires are able to use an entirely different form of echolocation, which seems to have developed quite independently of that already described. Whereas the sounds emitted by the other insect-eating bats leave the mouth, the horseshoe bats use the nostrils. These are situated in the center of the bowl-like horseshoe membranes, and behind them each horseshoe has a strangely shaped upright

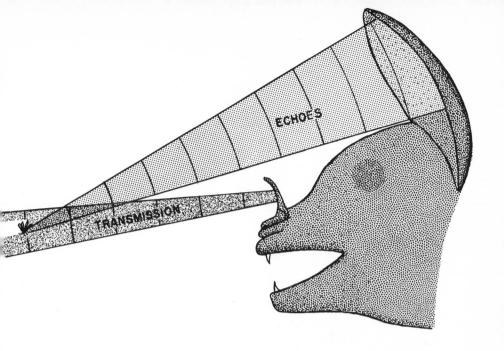

42. This photo shows how the vertical nasal appendage of some horseshoe bats and false vampire bats may concentrate sound transmissions. In these bats, the intensity of the echoes is the means for judging their distance from prey; not the time taken for an echo to return, as in some other bats.

extension that acts with the horseshoe to project the ultrasonic sounds forward in a narrow beam with astonishing accuracy and intensity.

There may be some control over this also, because the contours of the membranes can be varied by the movements of a complex system of muscles in them, and by chambers in the air passages that appear to have no other purpose than the amplification of the sounds by resonance. The upright extension of the horseshoe may also play an important part in directing the sounds away from the ears of the bat transmitting them so that they will not confuse returning echoes, but we cannot be sure of this. Such a method of direct sound projection, and a pair of very movable ears, give this kind of bat an echolocation system that is superior to those of other bats.

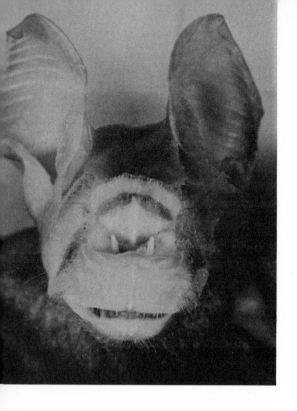

43. *Hipposideros diadema* is a leaf-nosed bat in which the strange nasal formation functions similarly to that of the horseshoe bats. The folds in the ears may be involved with directing sounds into the ear canals in a particular way. This bat is an intermediate type that cannot move the horseshoe or leaf, nor are the ears movable. There is only one leaf-nosed bat in North America, the large-eared bat, genus *Macrotus*, which is found from California to Mexico. But there are many in South America and the Old World.

The greatest difference in the two systems is that horseshoe bats emit pure-frequency sounds, each species having its own characteristic frequency, the difference between the lowest and the highest being as much as 25,000 cps. The pulses are longer than those emitted by the other insect-eating bats, too. Instead of a duration of one or two thousandths of a second, they are about a tenth of a second. This means that an echo has returned before the pulse has completed its transmission. However, horseshoe bats do not orient themselves by the time it takes the echo to return from an object, but by the change in intensity of the echo as the object is approached. These bats move their heads from side to side, and carry out fast and complicated movements of the ears, whereas the other insect-eating bats keep their ears motionless.

Because its ears are so movable, plugging one of them does not handicap the horseshoe bat to any extent, but plugging both ears or the nostrils does, of course. Horseshoe bats, as all bats, use cries for communication that are quite different from those they use for echolocation.

As with so many other groups of animals, the bats have produced some variable types. One, a fruit-eating bat known as *Rousettus*, which lives in caves, has developed an echolocation system like that of the insect-eating bats, in which judgment of distance is by the time taken for the echo to return to the animal's ears. However, being a fruit-eating bat, it has also retained useful vision and smell capability. With this unusual combination of characteristics we find also that the sounds it makes are not made by the larynx, but by the tongue. They are nevertheless just as effective.

Birds that use
echolocation

Most people have heard of the Chinese delicacy bird's-nest soup. The birds that build the nests from which this soup is made are among the very few that are nocturnal. They are found in Sarawak, and during the day they hide in the depths of dark caves. A species of swift (*Collocalia maxima*), they use a loud clicking sound for echolocation in the caves, where their eyes are useless. The clicks are of short duration but have mixed frequencies and are made in addition to normal bird sounds. This clicking seems to be synchronized with wing movements, but it can be turned on and off, so it is not quite so easily investigated as the sounds made by bats. It also appears to be used in courtship, further complicating its interpretation.

Another bird, the oilbird of Caripe (*Steatornis caripensis*), found in Venezuela, Peru, Brazil, Colombia, Ecuador, high in the Andes, and in Trinidad and Guyana, is also a nocturnal cave dweller that uses echolocation for finding its way around inside the deep caves. Its local name, guacharo, is Spanish for "one who cries and laments," perhaps because as these birds twist and turn in the dark they also emit shrieks and squawks. In some places this guacharo is mistakenly called an owl because of its nocturnal habits, and perhaps because it is about the size of a hawk. Its diet is fruit, and the kinds of fruit, palm and laurel, on which it feeds give it an exceedingly high body content of fat. This is the reason for its common name, oilbird.

Except when they turn in flight, oilbirds voice evenly spaced clicks or snapping sounds with a frequency of about 7,000 cps., and, like bats, if their ears are plugged they can no longer navigate in the dark. They collide with the cave walls and with each other. Like the pulses of the insect-eating bats, their clicking sounds vary in number or frequency as an object is approached.

Both these species of cave-dwelling birds have retinas that are highly specialized for dim-light vision, and, with excellent night vision, they use only their eyes for getting around in the dark outside their caves. Their pupils expand and contract very extensively, and they have a brilliant eyeshine when light is projected toward them.

Echolocation
under water

How Sound Travels in Water

Sound travels better in water than in air; it is louder and travels faster. The loudness is easy to confirm while lying in a fairly full tub of water, by covering one ear with a finger to keep out the water and putting that side of the head underwater, at the same time knocking on the bottom of the tub with the other hand. The sound will be quite loud, but barely audible if the head is raised out of the water. Even small explosions underwater can be picked up at considerable distances from their sources.

Sound also travels faster in water than in air, up to a mile a second, with slight variations for salinity, depth, pressure, and temperature. This is five times faster than in air, and so a sound wave of any given frequency has to be five times longer than it would be in air.

The humidity of air will vary the speed at which sound will travel through it. A sound with a frequency of 1,000 cps. travels 366 yards per second in air, so each cycle or wavelength is $\frac{366}{1,000}$ or 0.366 yards long. But in water the same sound would travel approximately five times as far, let us say 1 mile in a second. There would still be 1,000 cps., so each cycle or wavelength would have to be $\frac{1,830}{1,000}$ or 1.83 yards long.

88

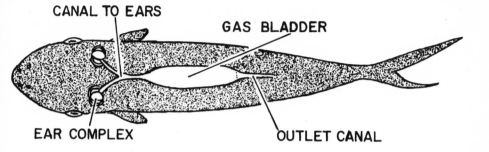

CANAL TO EARS

GAS BLADDER

EAR COMPLEX

OUTLET CANAL

44. In some fish, the gas bladder (often called the swim bladder) has direct bridges to the inner ears. This bladder can pick up vibrations, like a drum, and pass them to the ears.

The frequency would not change, only the wavelength.

How Fish Hear

Nature would hardly neglect such a valuable means of communication or recognition of danger, so we find that not only do fish make quite a variety of sounds, but they respond to a wide range of sounds also. Many people have the idea that because a fish has no outer or middle ears, it cannot hear. But it does have the important inner ear, and this responds to sounds conducted to it by the skeleton and skull. Many experiments have now confirmed this.

The inner ears are frequently embedded in the bones of the head in fish. There is also a direct canal or bridge between the inner ears and the gas bladder (sometimes mistakenly called a swim bladder) of perhaps five thousand species of the fish which have not discarded this chamber, and it has been shown that the gas bladder can act as an amplifier of sound, both transmitted and received. Sometimes the bridge between the gas bladder and the ears is bony, similar to the hammer and anvil in higher animals, but the result is the same.

We know definitely that the inner ear of the fish is not just a balancing organ or living gyroscope, as was once thought. It is the seat of most sound reception too, responding in many fish to sounds with frequencies up to 10,000 cps., and in a few to ranges between 45 and 13,000 cps. Not all the sounds are registered through the ears; the lower frequencies often reach the brain by way of the lateral-line system, only the higher frequencies are routed through the ears. At least one claim has been made that fish can respond to ultrasonic vibrations, but there is no certainty of this so far.

What seems to confirm that fish can recognize, and are sensitive to, a wide range of sounds is the size and development of the hearing centers in their brains. These have now been checked in many species.

Some fishermen believe they can attract fish with sound, but there is no proof of this. Fish are quite as likely to be scared off by it. Sharks, however, certainly can be attracted in this way, and it seems that large snappers and gropers are, too. But the sound must be made underwater, and must be of considerable volume. So far as sharks are concerned, their skeletons being of cartilage instead of bone, the range of sounds they hear most effectively is that which is received through the lateral line system, and this means sounds in the lower-frequency ranges.

THE DOLPHIN'S TRACKING POWERS

Because of its possible application to undersea warfare, all the principles of echolocation used by marine mammals have been studied intensively, and of all those animals that have been experimented with, the dolphin and porpoise have proved the most cooperative and the most informative. The range of sounds that dolphins can hear may not be as great as that of bats, but according to some tests, dolphins do respond underwater to frequencies as high as 170,000 cps. Furthermore, it is possible that a dolphin's brain is even more discriminating than that of any bat, for it has to unscramble echoes from objects, from other living creatures, from the bottom and the surface of the ocean, from water flow around its own body, from waves on the surface, and from ship machinery, all of which have to be decoded, analyzed, and identified.

Apart from the sounds we can hear the dolphin make above water, it appears to have three different and independently controlled kinds of sounds for underwater communication and echolocation. It uses squeaks and whistles for communication with its own kind, but its echolocation sounds are clicks that are so versatile they can be transmitted one at a time or up to 800 a second, the duration of the shortest being about a thousandth of a second. In pitch, the sounds vary from 20 to 170,000 cps. Low-frequency sounds travel farther than high-frequency, so the latter are used more in short-range identification of prey and obstacles in the dark.

45. The dolphin transmits patterns of clicks, which are reflected from any unseen object just like the sonar of a submarine or ship. The reflected echoes are picked up by the dolphin so that it can "home in" on the object or fish.

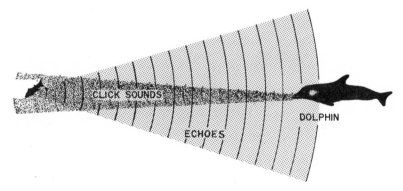

CLICK SOUNDS

ECHOES

DOLPHIN

This animal's powers of interpretation enable it to identify both the size and the species of fish from which echoes are bouncing. In fact, tests have shown that even in utter darkness it can differentiate between two objects in which there is only a quarter of an inch size difference, and it can detect an object no larger than a sixth of an inch in diameter. It will always avoid collision with an invisible glass or plastic sheet placed in its path, and it cannot be confused by man-made sounds, or even recordings of its own sounds retransmitted to it.

While just cruising around, the dolphin sends out bursts of sound about every twenty seconds. If an echo results from one of these, the dolphin then transmits continuous sound to identify the source. If this is food, it moves in, scanning its sound beam from side to side within a ten-degree arc until it reaches the source. Any surface splash will start off a burst of sound and exploratory behavior.

What is so remarkable about all this is that the dolphin's ear is quite closed, and the eardrum and acoustic bones are loose. It must therefore hear in a different manner from that of land animals. The sound reaches its middle ear through the body tissues and skull, which means that, although the animal hears well in water, it cannot hear so well in air. The middle and inner ears are a single complex, the acoustic nerve is immense, and the whole mechanism has reached a higher level of development than in any other animal. The hearing centers in the brain are four times the size of the visual centers, the opposite of most land animals.

46. Although airborne sounds are not so audible to them as those in water, trained dolphins can be summoned by a whistle or bell if it is fairly close to the water. Even in complete darkness they will home in accurately on a noise or an object revealed by their own sonar transmissions. The two shown here perform in answer to a whistled command, which reaches them through the water.

How is such accurate acoustic interpretation accomplished? It seems incredible to us; yet our visual judgment and interpretation are just as complex. Distance, size, color, shape, form, brightness, and other qualities all seem simple things for us to judge. We can concentrate on a small object like a grain of sand and ignore everything else around that falls within the visual images in our eyes. So, in just the same way, the dolphin ignores all sounds but the echo on which it is homing in. If, when we are looking at the grain of sand, something moves suddenly to one side of us, we are immediately alerted and can shift our area of concentration. The dolphin can do this too, when sounds change or new ones reach it.

Differently shaped and sized objects appear so to us because the light rays reflected from them produce different images in our eyes. They produce different echo patterns (sound images) in the dolphin's hearing centers. The dolphin's greatly developed hearing sense is doubly necessary to it because it has no sense of smell, not even an olfactory center in the brain. This develops normally before birth, and the olfactory nerves are well formed when the embryo is no more than half an inch long, but they all disappear before it reaches a length of two inches. Not only does the dolphin have no sense of smell, but what is even more extraordinary, it has no vocal chords. So how does it make sounds? In answering this we are on less certain ground.

Some scientists believe that the sounds are made by forcing air from small sacs close to the blowhole, the sounds passing through and perhaps resonating in a fatty melonlike bulk of tissue in front of them, because, while the animal is submerged, the blowhole must be tightly closed. But others believe that small pieces of cartilage in the air passages may play a part in sound production. The facts still have to be established, except

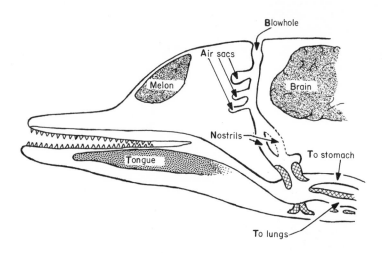

47. The dolphin's breathing apertures have fused together into one "blowhole," and somewhere in this system it produces its clicking sounds. The small cross-hatched patches represent cartilage.

that the mouth moves when sounds are made on the surface, and some sounds do seem to come from the blowhole when this is open above water.

Echolocation by Whales

Even on a brilliant day there is no appreciable light in the oceans below a thousand feet, so how do other marine mammals, which wander down as far as four thousand feet, manage to locate prey? Again it is with echolocation. The killer whale (*Orcinus orca*), like the dolphin, produces high-frequency sounds and clicks ranging from 50 cps. to perhaps 80,000 cps. with which it navigates and avoids obstacles. These sounds travel great distances. In this animal, the sounds also differ in the sexes.

These so-called whales are, in fact, giant members of the dolphin family (Delphinidae), and evidence of their ability to

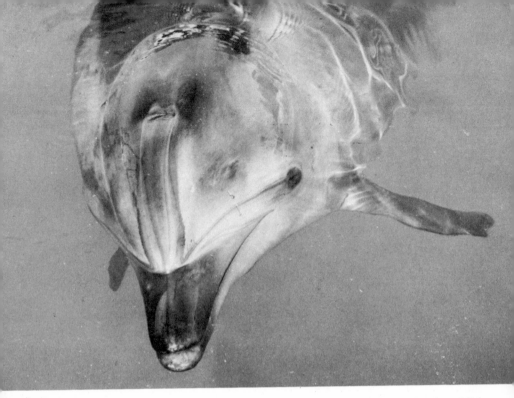

48. The dolphin's blowhole at the top of the head is clearly visible here. It is closed until the animal reaches the surface, so underwater sounds must be transmitted by some other way.

find prey in the dark depths can always be found in their stomachs when they are killed; one killer whale can have as many as three dozen seals and porpoises in its stomach at one time.

Instead of the whistlelike call of the dolphin, this animal seems to use rapidly repeated grating screams for communication, using clicks at the rate of from 500 to 2,000 a second for echolocation. True whales also produce pulsed sounds for echolocation. When cruising, they seem to pay most attention to the low-frequency echoes, which will give long-distance information, but when closing in on prey, they respond to higher

49. The skull of the dolphin from above and below. (Top): The depression beneath the blowhole that holds the air sacs can be seen. (Lower): The bones that support the laryngeal tissues can be seen beneath the skull.

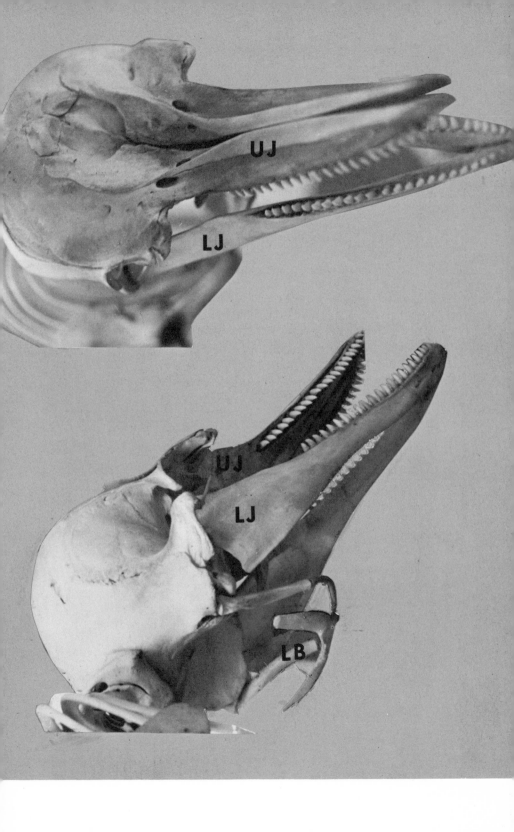

and higher frequencies, and thus establish the identity of what they are approaching.

There are reports that when sounds have been made to call trained dolphins during experiments, sea lions have responded as well, and from considerable distances. There is every indication that these animals and seals also use pulse-modulated echolocation for hunting fish in darkness. The loss of their eyes appears to make no difference whatever to them, but their vision is used and is quite efficient when there is light to see by. As the light fades, their echolocation is intensified, and some of them feed through the night on squid and other fish that rise from the depths of the ocean where they stay during the day. Little is known about the navigation of the dugong and the manatee, but their hearing is so acute that hunting them is exceedingly difficult.

EXPERIMENTS ON SHARKS

Sharks have good hearing for a certain narrow range of tones. Audiograms (recordings of response to sound) have been made on them by a number of scientists, and these show that in some sharks this range is from 100 to 1,500 cps., and their greatest sensitivity is in the band from 400 to 600 cps. This range is evidently below that of most of the noises in a shark's immediate environment, and can therefore be heard without interference from other sounds, enabling the shark to locate a sound source accurately and move toward it. Some species are responsive to even lower frequencies, 20 to 60 cps., which are audible over considerable distances, but these may be received only through the lateral-line system.

The frequencies to which sharks respond most will be found among the sounds made by a fish struggling at the end of a line, or by people swimming, especially those who are tired or in distress, and it is possible that such sounds can be traced from considerable distances. Darkness is therefore no protection against questing sharks.

The true hearing capabilities of sharks need more investigation, because some of the experimental findings so far differ widely from each other. Though species differences are to be expected, the cartilaginous skeletons of all sharks place limitations on the range of wavelengths that will reach their inner ears in useful volume, and claims of up to 7,000 cps. seem rather high. However, we may be underestimating a shark's powers, the full range of which has probably not been uncovered. Its ability to recognize the direction from which sounds are coming is good, and this has not been satisfactorily explained yet.

Other strange ears

Most animals that hunt or forage at night have large ears, which usually accompany sensitive hearing centers in the brain. Just how sensitive can these be? Well, the aardvark or earth pig found in Africa, *Orycteropus afer*, for all its 150-pound weight, lives on ants. It has long ears and a blunt long nose, which make it appear as though someone had crossed a donkey with a rabbit, and then crossed the result with a pig. It is quite an ugly animal, but a very efficient one. Those long ears that can twitch around in many directions can hear the movement of termites in a mound, and in the dead of night the aardvark listens for these and digs them out, consuming them in vast quantities.

There are other nocturnal animals that are able to do this. The aye-aye (*Daubentonia madagascariensis*) (Figure 32) listens for the larvae of wood-boring beetles, and needles them out with a very thin middle finger on each hand. Even stranger is the bat-eared fox (*Otocyon megalotis*) (Figure 31) in Africa, which also feeds on termites and other insects, with occasional fruit or small vertebrates. It, too, has ears that are each almost as large as its head. The fennec fox (*Fennecus zerda*) of northern Africa, with equally large ears, is, on the other hand, a real hunter, listening in the dark for rats, birds, lizards, or insects, alert to their slightest movement, even the breathing of some of them.

100

UNDERGROUND BURROWERS

Animals that live continually underground like the mole, and even some burrowers that leave their burrows during the night, have ears that are barely visible, perhaps just small apertures with no pinnas; or with the apertures covered by fur. These are adaptations to prevent dirt in a burrow from getting into and blocking the ears. There is some disadvantage in this, of course; but it may be compensated by picking up low-frequency vibrations which travel from the ground through the bones and skull direct to the inner ear, as in fish.

There must be less discrimination with this kind of hearing in mammals, but it is evidently quite adequate for their modes of life. Moles, in fact, may use not only perceived vibrations for gathering information, but they may also use a simple form of echolocation, because they make twittering sounds when testing their surroundings, and this could certainly be an excellent way to discover the distance from where they are to the end of a burrow or to a chamber in it.

The burrowing lizards and all snakes have only an inner ear. This also responds directly to ground vibrations, which pass to it through the skeleton and skull, again somewhat like that of fish. There are also nonburrowing lizards that have no eardrums; but eardrums are always present in species that communicate by uttering sounds. The eardrums are quite well developed in geckos, which, as well as being nocturnal, chatter in a simple way. And the crocodiles, which protect their ears with movable flaps while underwater, certainly hear frequencies from 50 to 4,000 cps. They also communicate with sound.

On the whole, probably all vertebrates can detect sound vibrations through the air, the ground, or water, and there are many that have a better developed hearing apparatus than our own.

Hunting by smell

Like hearing and vision, smell and taste can be developed for use in the dark. Smell and taste are very similar processes in man, and probably in most land vertebrates; they both function by dissolving chemical particles in a gland solution, the saliva in the mouth, and a mucous secretion in the nose. Animals living in water do not need such glands; the particles are dissolved in the water that passes through their mouths and nostrils. There are exceptions, like the dolphins, which have no nasal or olfactory function at all.

WHAT SMELL IS

An odor gives information on something that may or may not be within visible range. It may be close or it may be some distance away, whereas taste is usually possible only at close range. Some fish and most of the reptiles are exceptional in this, but for the majority of vertebrates, it probably remains true. Most animals, therefore, find and test their food by scent rather than taste. If their nostrils become blocked, some of them might even starve.

Man's sense of smell rapidly becomes satiated so that he does not remain aware of a smell after a short time. That is why people in a smoke-laden room are less aware of the atmosphere than anyone just entering from the fresh air outside. How ani-

102

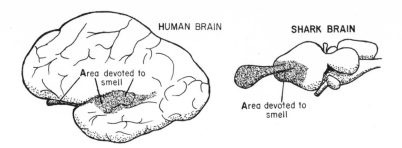

HUMAN BRAIN   SHARK BRAIN

Area devoted to smell

Area devoted to smell

50. The human brain devotes a relatively small amount of its volume to the sense of smell, and much more to vision and comprehension. The shark brain, on the other hand, uses a much greater part of its volume for smell, which is one of the predominant factors in its awareness. The exact amount is not yet known, but it probably approximates the shaded area shown.

51. A gorilla's head cut away to show the smell mechanism. On each side, the nostril opens into a nasal cavity. This is lined with a moist membrane that catches and dissolves the particles of scent in the air. The solution stimulates fine nerve endings, which send electrical messages to the olfactory lobe or bulb. This is an extension of the brain just as the eye is, and it is connected with other parts of the brain devoted to memory and the coordination of all the senses, so that what is smelled can be identified and related to what is seen, felt, heard, or tasted.

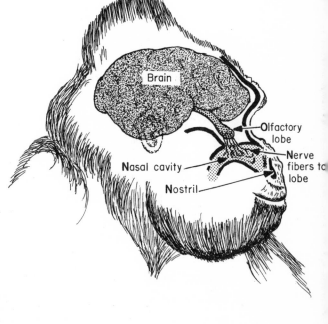

Brain

Olfactory lobe

Nerve fibers to lobe

Nasal cavity

Nostril

mals that depend on their sense of smell overcome this we are not sure, but it may not be a great handicap to them; fortunately, even in man, being satiated by one kind of smell does not reduce sensitivity to others.

In those animals that rely on their sense of smell for their survival, smell actually controls much of their behavior. Their association of smell with other sensations, such as vision and taste, is the basis on which they develop all their exploring patterns.

MORE EXPERIMENTS ON SHARKS

Most sharks have a very acute sense of smell, especially for blood; and many electroencephalograms (recorded wave patterns of brain activity) have been taken from sharks in studies of their sense of smell, their olfactory sense. Like most primitive animals, the sharks have very large areas of their brains devoted

52. The head of a hammerhead shark showing the nasal groove. With a sensitive and broad smell receptor area like this on each side of its wide head, it has an astonishing tracking ability, unequaled by other sharks.

to the interpretation of smells, and in many species the olfactory apparatus occupies two thirds of the entire brain. This is why sharks can not only smell from great distances, they can also smell very faint odors, fractions of one part in a million of water. So much do they rely upon this sense that removal of the olfactory lobes from the brain causes them to develop the same delayed reactions and lack of spontaneous action that higher animals demonstrate when their whole cerebral hemispheres (main lobes of the brain) are removed.

Much of this has long been known, but making the encephalograms also refines our knowledge by telling us which smells affect sharks most. Some of the results, which may be important to fishermen, show that, apart from blood, tuna flesh has exceptional shark-attracting power.

Direction From Smell

Just as the dolphin moves its head from side to side to judge the exact direction of sounds and echoes, so a shark moves from side to side to identify the exact direction to the source of a smell. The two nostrils analyze different densities of odor, just as the ears do with sound, the animal always turning toward the side from which the stronger scent comes. Plugging one nostril will make a shark swim in circles. The lemon shark does not appear to detect food until it is within visual distance of it, and this has made some experts doubt that sharks are so keen-scented, but the experimental findings on other sharks seem to be quite irrefutable.

The wider apart two nostrils are, the better equipped is an animal to find a food source in the dark. Hammerhead sharks are especially well equipped for this. On the forward edge of the wide head there are a pair of olfactory grooves and concave slit nostrils that are unusually sensitive; added to the questing movement from side to side, they give this shark a formidable hunting power.

The superior power of wide separation between a pair of organs applies equally to ears. Eyes that are wide apart have the advantage of better judgment of the distance to anything they fix on; better stereopsis. The bony fish do not have this advantage of the hammerhead shark and, except in the most predatory kinds, they do not seem to have the acuteness of smell of the sharks as a whole.

SMELL ORGANS IN FISH

Whereas our nostrils lead into the air passages, in fish they are merely inlets to sacs or chambers lined with odor-sensitive cells, and out of which there are also small outlet nostrils. This arrangement ensures a constant flow of water over the sensitive cells, perhaps a more effective system than ours, in which we have to sniff or breathe in and out rapidly to detect some odors. The largest chambers are found in eels, which are therefore well equipped for hunting at night. The moray eel certainly excels at this.

When bony fish become distressed, as when they are caught on a line or in a net, they give off a skin secretion that has a very easily detected smell. It is thought that this may play a part in warning other members of its species, which can then move rapidly out of the danger zone; but whatever its true purpose, it certainly has the power to attract predatory species to the spot, and this applies especially to sharks. A small amount of water taken from the area of a distressed fish and poured into a shark pool will immediately make the sharks restless, so that they will start to hunt around. But this may not happen every time, so we still do not know everything about this secretion and the response to it.

53. The moray eel hunts at night by smell. Two projecting nostril tubes can be seen above the mouth, and these lead to large nasal sacs which have outlets higher up, by the eyes.

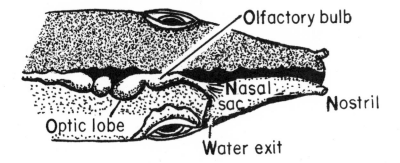

54. The large nasal sacs, or chambers, in the noses of the moray eels (family Muraenidae) take in water through nostrils at the front and pass it out again through other nostrils just by the eyes; liberal nerve trunks connect these sacs with the olfactory centers of the brain. The cut-away diagram shows the position of the right olfactory center.

TRACKING ON LAND

The way land animals track by scent is well known. Dogs, especially certain breeds, are examples always around us. But many animals have equally effective, and even better, olfactory

55. The long beak of the nocturnal kiwi (*Apteryx mantelli*) has nostrils at the very tip, which probably are particularly sensitive to the odor of worm slime, for the bird plunges its beak into the ground to capture large quantities of worms.

apparatus than that of a dog. Some are sensitive mainly to odors given off by their prey. The mole, for instance, seems to be most sensitive to the smell of worm slime; it will eat other things, but its preference is for worms, and it finds them in the dark.

Worms are the diet of one of New Zealand's flightless birds, also. The kiwi is nocturnal, being most active at the time its chosen prey is coming above ground. Though a worm may stay below ground during the day to escape the many species of bird, amphibian, reptile, and other animals to which it is a delicacy, it has no escape from the kiwi. This bird's long beak has nostrils at the very tip, and when there are no worms on the surface, this long probe is thrust into the earth, where it detects them by their odor just as the mole does, grasping and withdrawing them deftly. Except for this, it would be much more beneficial to the kiwi to have its nostrils higher up the beak as most other birds do.

In mammals, the longer or more prominent the snout, the more likely is the animal to rely on its sense of smell to a major extent. Frequently, when the snout is short, the animal relies more on its eyes or ears, but most likely its eyes.

Apart from the elephant, one of the longest snouts is possessed by the giant anteater (*Myrmecophaga tridactyla*) of Central and South America. This is another animal that lives on termites; but although its ears are quite sensitive, it does not depend on them to the same extent as the three animals in Figure 31. Instead, it detects its prey by odor; then it rips open the termite nest with giant claws and with its sticky, prehensile tongue it scoops up the insects in large numbers.

56. One of the longest snouts seen in the animal world belongs to the giant anteater (*Myrmecophaga tridactyla*). This snout has two basic functions. In addition to housing a very sensitive odor-detecting membrane, it is useful for thrusting into small holes, where termites can be swept up with its long, sticky tongue. These animals, which live in Central and South America, are large and powerful.

This animal is not essentially nocturnal. It is arhythmic, functioning at any time. Other excessively developed snouts are found in the Cuban solenodon *(Atopogale cubana)*, occuring in the West Indies, and in some of the shrews.

There are a few animals that function well at night without leaning more heavily on one particular sense. Instead, all three senses of vision, hearing, and smell are well balanced. The raccoon is an excellent example of this all-around efficiency. Although this animal will be seen foraging in the daytime, its greatest activity is at night. When the night is really dark, it will rely on its nose and ears for information; but if there is a moon, its eyes serve it better than ours can serve us.

Taste in the night

Although smell and taste are such similar senses that sometimes it is difficult to consider them separately, in fish the organs that serve these senses can be more widely separated in their location on the body, and so the senses themselves can be more individual. The nasal organ is always to the front of the head, but taste receptors (buds) are found outside a fish's mouth as well as inside. In fact, there is some doubt that there are any at all inside the mouths of some species. They may be on the skin of the lips, along the body, on the head, or even on the fins.

BARBELS AND OTHER TASTE ORGANS

Special taste organs known as barbels, whiskers, or feelers are carried by some fish like external tongues, and these brush against things and recognize them by taste, just as the nose detects and recognizes them by smell. Touch receptors on the barbels may also play some part in their sensitivity, but the use of the taste organs in muddy water or darkness may be almost as effective as a sense of smell in some species of fish.

The bullhead catfish (*Ictalurus*) finds food from quite a distance by taste alone if the current carries the clues toward it. If the smell nerves become damaged this ability is not reduced at all, but it is if the taste nerve is damaged. If there is no current, this fish swims from side to side as it goes forward, testing the

111

area thoroughly with the thousands of taste buds on its barbels and body.

It is possible that the function of these external taste buds is only to detect food, and not to identify what kind of food it is, nor to make any sort of discrimination. The fish has

58. The carpet shark (*Orectolobus ornatus*) has taste barbels around its mouth. It also has many weedlike appendages, but to what extent these may be sensitive to taste is not known. They may be related more to camouflage.

57. (Top): The two barbels seen in the sturgeon (*Acipenser fulvescens*) enable it to detect food in dark and turbid water. Catfish are similarly armed with tasting organs. *Pilodictus* spp., which occurs in North America (center), has particularly long ones, and those of the long-tailed catfish (*Euristhmus lepturus*) of Australia are much fleshier.

many other taste buds in the mouth, throat, and gill chambers, and it is these that appear to prompt it to reject unpalatable food it has picked up. However, some other fish appear able to discriminate between different foods with the barbel taste buds, and many, unlike the bullhead catfish, are said to have no buds inside the mouth.

In freshwater fish, these internal taste buds are more plentiful in the gill chambers than they are in marine fish. They are arranged along the front surfaces of the gill rakers, so they are close to food taken into the mouth, and also so situated that the water passed through the gills flows over them. These taste buds vary greatly in size, getting larger as the fish grows, but we do not know whether the size of the taste buds affects their taste discrimination.

TASTING THE AIR

In many American Western films, sooner or later an Indian will tell a white man that he speaks "with forked tongue." Evidently Indians associate the forked tongue with double-talk, lying, or treachery. Poisonous reptiles, and many others that actually have forked tongues, are also looked upon as treacherous, a word often wrongly used for "dangerous," for some people still think that the forked tongue of a snake is used to inject its venom.

The forked tongue, like everything else in nature, has a definite purpose, and it may well be the most useful organ of any animal possessing one, especially in the dark. Most of us have heard that when a snake is killed, its mate often finds its way to

59. Shown here are a small python's forked tongue leaving its sheath; and the pits leading to Jacobson's organ, into which the tongue flicks when tasting the air.

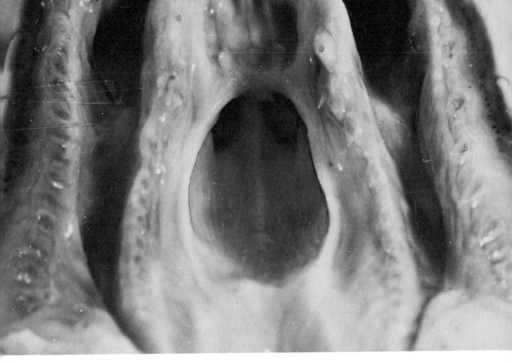

the spot and lies in wait for the killer, eventually avenging itself on him. This is quite fallacious, and, though it may appear
to have happened at times, it must have been purely coincidental.
The facts are quite different.

JACOBSON'S ORGAN IN SNAKES

The relationship between the forked tongue and this story is
to be found in the tongue's habit of flicking out of the mouth.
This way it picks up particles in the air, which it flips rapidly
back into a pair of crypts or cavities in the roof of the mouth,
called *Jacobson's organ* or the *vomeronasal organ*. These crypts
open into hollow cavities lined with sensory cells similar to those
in the nose, and they are connected with the brain by numerous
branches of the olfactory nerve. These cavities are highly sensitive taste areas, which can identify anything by only a few
molecules that may be floating in the air, and they give many
snakes a tracking power that is far superior to that of a dog, or
perhaps any other animal. Such tracking power is essential to a

60. Jacobson's organ is separate from the nose, its access being only
through the roof of the mouth. It is served through a canal from
Harder's gland around the eye, which provides its lubricant.

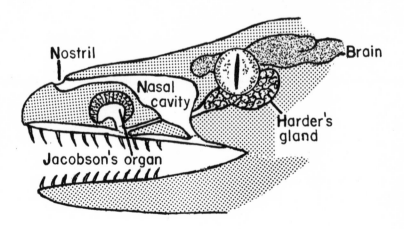

61. *Megalania prisca*, a giant prehistoric lizard (reconstructed) that appears to have had the Jacobson's organ. This seems to indicate that this organ was a very early reptilian adaptation.

reptile that may bite prey in the dark, which may then escape to die later from the effects of venom, perhaps in some remote and hidden place, where the snake would otherwise never find it.

This is a reasonable explanation for the use of such an unusual organ so far as certain snakes are concerned, but it would seem to be less applicable to those snakes that hang on to their prey, or to lizards and chelonians, which also have Jacobson's organs. It has been suggested therefore that the prime function of the organ is in the identification of food through taste, even though the impression some tests produce is that it picks up scent particles. But taste and smell being so similar, any great distinction seems irrelevant, and we can still say that Jacobson's organ is a highly sensitive taste area that can identify immeasurably faint trails without in any way reducing the usefulness of the nose.

When an animal such as a frog or mouse passes through the grass, it leaves solid particles of a substance from its skin on everything it touches. Many of these float off into the air, and a snake picks them up by constantly flicking its tongue out and back into Jacobson's organ. This amplifies the effect on its senses much more than if the nose alone were used, and it may be because of this amplification that the animals can use it for sex recognition. Blindfolded or blind snakes will court each other, but never if the nose and Jacobson's organs are blocked.

All sense organs need some kind of lubrication to function well; the nose, the ear, the eye, the tongue, all are kept moist by gland secretions, and Jacobson's organ is no exception. The interesting feature, however, is that in snakes and lizards this organ uses the glands that lubricate the eye, one of which in higher animals produces tears. Canals have developed from these glands to Jacobson's organ.

This organ is no recent innovation. The small bone that protects it is found in the fossils of many long extinct reptiles. It has been retained in functional form only by reptiles. It seems possible that, together with the forked tongue, it was possessed by the giant reptiles, for *Megalania*, which was related to the goanna types and grew to a length of 20 feet, appears to have had it. Not only is it found in the embryos of a number of higher animals, disappearing before birth, but vestiges of it still appear occasionally in the mouths of certain modern mammals, particularly in the rat and the guinea pig.

Jacobson's organ certainly seems to be associated with a nocturnal life, when eyes are at a definite disadvantage. Its absence in turtles, which have never been very active at night except for the secretive laying of eggs, appears to support this. But crocodiles are nocturnal, and it is absent in them, too. This supports an alternative suggestion, that the organ may be less useful underwater and was therefore discarded by the ancestors of these animals. On the other hand, the sea snakes have retained it, seemingly as functional as in the land snakes, and they never come ashore.

When a snake is preparing to shed its skin, it becomes very inactive. This has always been considered to be because the eye coverings, which are shed with the skin, become somewhat opaque (milky), so the animal is unable to see. But there is another factor that, added to this, puts the snake at a double disadvantage. The covering of the tongue is shed also; while this is degenerating and hardening prior to the molt, it is undoubtedly much less efficient in its collecting of particles for Jacobson's organ, so the snake is less active at night as well.

When it is examined under a microscope, we find that Jacobson's organ is lined with a sensitive membrane like that of the nose, and this is kept lubricated by the secretion from the eye glands (Harder's gland), just as the mouth is kept moist by the salivary glands in all animals. And it is possible that the secretion used by Jacobson's organ adds to the lubrication of the mouth in those venomous snakes in which the true salivary glands have evolved into the venom glands.

There must be a strong relationship between the function of Jacobson's organ and the actual food needs of a snake, because tests have shown that in newborn snakes which have never previously been fed, the tongue flicking starts and attack postures result in the presence of skin substances of small animals dissolved in water. What is even more remarkable is that the preferences of adult snakes for particular kinds of prey are shown also by the newborn snakes when they are presented with various solutions of skin extracts. Theoretically then, a newborn snake can hunt or detect the desired kind of food in the dark right from the start.

Touch and
special senses

With the exception of certain fish, the sense of touch will seldom aid an animal in tracing food, but it does have a distinct role in helping to avoid unseen danger. The cat's whiskers (*vibrissae*) are special touch organs. They are the same width as the animal's body, and this enables it to test openings in the dark well enough to know if its body will pass through. Many other animals, rodents particularly, have similarly sensitive whiskers.

The mole is well equipped in this way, and besides a complete ring of whiskers around the nose, it has a collection of special little clusters of nerve endings called *Eimer's organs* on the end of the nose. These are so carefully arranged that they may be able to combine with the vibrissae to identify the direction of air currents and compression waves created when the air it pushes along its tunnel meets obstacles. It uses them to test the air outside its burrow if it emerges at night.

The star-nosed mole (*Condylura cristata*) is especially well equipped for finding its food with ultrasensitive tactile organs which may, in fact, also have some taste function. Around the end of its snout, it has twenty-two small, naked, fleshy appendages arranged like a star, and these appear to be used for detecting food in swamps and at the bottoms of lakes and streams. If they were purely tactile they would not have to be naked,

120

but as taste organs they would, so it does seem that they perform two functions.

In fish, the touch receptors seem to be distributed most thickly in the same areas as taste receptors—over the head and in any barbels that may be present, and sparsely scattered over the body. It is often difficult to differentiate between these receptors and the function of some of the sensitive cells in the lateral line system, a complex of tiny organs usually in a line along each side of the body; but there are subtle differences.

Other receptors that respond to temperature change can be important to some fish that tend to move from place to place in order to remain in a steady temperature rather than at a particular depth. Some of these can detect a temperature change of as little as a thirtieth of a degree centigrade, and, though gradual temperature change will not necessarily harm fish in the adult stage, it is very important that they spawn in an area of unvarying temperature, because the fry frequently succumb in any rise or fall.

THE LATERAL LINE SYSTEM

Sometimes the complex of minute organs that make up the lateral line system in fish is a line of individual pores. Sometimes it is a line of pores opening into a canal that runs the length of the body beneath the surface and within which are located the sensitive cells that record the disturbances around the fish. Frequently they are also grouped in large numbers in other areas such as the head, or around the tail. But wherever they are located, they are an incomparable means of sensing the unseen world around.

Some of these organs are very similar to the ear mechanism in what they achieve. Because they provide what may be con-

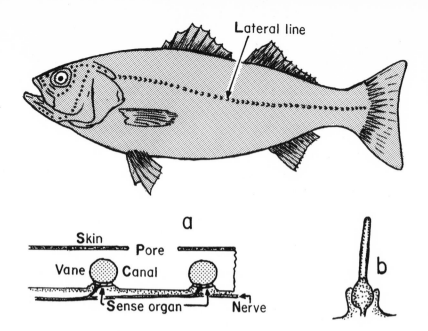

62. The lateral line system of a fish is not always just a line. The organs can be scattered, or in orderly formations over the head or elsewhere. Two lateral line systems are shown beneath the fish: (*a*) is a canal with pores to the outside, and sensitive organs (vanes) inside the canal, each organ being served by a branch of a nerve running with the canal; (*b*) shows, highly magnified, an organ which extends out beyond the body surface.

sidered a compromise between the senses of hearing and touch, they are also called an *acoustico-lateralis system*. In fact, these organs and the ears often develop from the same common source in the embryo. Some scientists even believe that the ears evolved from the lateral line system. But the impulses sent by the lateral line organs to the brain along their nerve fibers can be in response to low-frequency sound vibrations, minute electrical currents, body movement, or water pressure. It is not always clear, however, if a fish is responding to minute water currents created by another fish's movements, the sounds created by those movements, or the electrical currents transmitted by the other fish's muscle action. Temperature is involved too, because it varies the effectiveness of these sensations.

The first primitive vertebrates had lateral line systems, and the invertebrate ancestors may have had them too. Some of the surviving primitive fish have lateral line patterns that could be similar to the very early ones, and in fact we find that the more primitive a fish, the more complex is the arrangement of its lateral line system. The bowfin *(Amia calva)* has nearly four thousand sensitive organs on its head, and the garfish (*Lepisosteus osseus*) has a similarly complex arrangement.

Fish that are active swimmers have some of the best lateral line systems, and those that are fast swimmers have them concentrated more toward the tail, which suggests that these systems also respond to the sensation of speed. Sluggish swimmers are more like the bowfin, itself rather sluggish, in that the lateral line organs are concentrated more to the front of the fish, and this could be related to an awareness of current speed and the surroundings.

There seems no doubt that a lateral line system aids fish in keeping in shoals and formations without touching when they cannot see each other. During the day, vision may play a large part in this, but at night each fish will respond to the pressure caused by another fish moving toward it, and the fish will all move in unison.

Experiments have been carried out on fish in which the lateral line organs are confined mostly to the head. Having been fitted with plastic hoods that left the mouth, eyes, gills, and fins free but covered the rest of the head, these fish were quite unable to detect vibration in the water, could even be touched, and they could not avoid one another. This confirmed other evidence that fish use their lateral line systems to swim in shoals and to change course in a compact mass without touching one another. This tight shoaling, which is a habit of some four thousand species, evidently reduces the risk of individual fish being taken by

predators; in the dark, hunting fish might sense the shoal as one large body rather than a collection of small ones.

Lateral line organs would be much less effective in air, and there is nothing like them now among land vertebrates, except in a few larval amphibians that live in water. Even those animals, such as the whales and dolphins, that have returned to live in the sea have nothing like them. Instead, they have developed their ears and the echolocation system to find prey and avoid obstacles in the dark.

The amphibians that have something of a lateral line system are mainly those living in lightless caves. Salamander larvae have it in functional form, and some even retain it into adult life, but in most instances it degenerates. The sense organs are usually found in rows on the head and body as in fish, but they do not seem so sensitive as in fish. They function only in the water and respond only to low-frequency vibrations.

Using Electrical Currents

Some lateral line organs pick up weak but very informative electrical currents in the water, and it seems that this perception of electrical phenomena is very similar to that for temperature and current sensing, and is in the same system. Many fish, especially sharks and rays, are so sensitive to weak alternating electrical fields that an artifically created electrical current will affect their breathing rate. The very weakest of electrical stimuli slows the heartbeat of some of them. So sensitive are these fish that they can be aware of one hundred-thousandth of a volt, and this is

63. The mulloway (*Sciaena antarctica*) has some prominent and brightly reflecting pits along its lateral line; photo shows clearly this organ's location and direction. (Below): The lamprey (*Mordacia mordax*) has sensitive lateral line organs encircling its head; view is of the underside, with the mouth (*M*) closed over the rasping teeth.

lower than the current generated by the muscles that move the gills of a fish, a level, however, that has been detected and measured by instruments. Apparently it requires only one ten-thousandth of a volt to affect the breathing rate of some rays.

Careful experiments have shown that not only are sharks and rays sensitive to these electrical fields, but they are also able to locate their source. One of the skates (*Raja clavata*) can detect the presence of animals buried in the sand solely by the infinitesimal electrical currents created by their gill muscles. Human muscles produce similar minute electrical currents, and the electrocardiogram (a record of heart action) is made from these. So the sharks' detection of swimmers in the dark may be made more certain by such currents.

A shark's lateral line system, which is so sensitive to these small electrical currents, is the kind in which small pores in the skin lead to a fluid-filled canal where the sensitive cells are arranged with tiny hairlike cilia projecting into the fluid. The canal divides close to the eye and passes both above and below it, perhaps giving double sensitivity to the region of the head. And perhaps it is this region that is most sensitive to pressure waves; sharks appear to turn toward and approach depth-charged areas.

In the Ocean Depths

The effectiveness of the lateral line system does not appear to be disturbed by depth. We find it used in both surface and deep-sea fish. Some relatives of the cods, known as rattails, are among the commonest of all deep-sea bottom-living fish, and they have heavily armored heads that are pitted with prominent canals containing long-distance lateral line sense organs.

Although rattails have what are probably the most sensitive eyes to be found in the fish world, most of the time they live below the level where light can penetrate except in the faintest degree. It is possible that the lateral line system is used by the rattails to obtain most of their food, picking up the disturbances produced either by tiny water currents or electrical currents, or perhaps even both, caused by the movements of other fish nearby.

This is supported by two facts. First, the water has no measurable movement of its own at great depth; and second, the sensitivity of the lateral line organs of many deep-sea fish will detect another fish moving its jaws and passing water through its gills, or eating, or swimming at a distance of anything up to sixty feet. It seems that when the sensitive organs project like little cushions of cells above the skin, each having a tiny cilium or filament, they register low-frequency vibrations and current pressures caused by movement; those that respond to electrical currents are probably located in canals.

Sound waves created by a fish moving its tail or fins travel very fast through water; nearly 5,000 feet a second, and this means that the low-frequency vibrations picked up by a fish's lateral line system give it an instantaneous awareness of another fish's presence or approach. There is another possible use for this system, somewhat like the echolocation system of the bat. The movement of the fish itself in the water sets up little sound waves that move out in all directions, and when they meet an object or another fish they are reflected back, the lateral line system recording them as well as the fish's own movement and speed.

In spite of examples such as that of the rattail, it is perhaps natural that fish with the least sensitive eyes should be equipped with very sensitive lateral line systems. Fish that are known to be blind have developed them to a high degree of efficiency. But all fish, even those that live near the surface, rely on their lateral line systems at night, and just as they will be aware of the movements of other fish in this way, so will they also be aware of the movement of an angler in a boat. This applies most to aluminum boats, because these drum and vibrate sounds through water very noisily. It is even possible that river fish can feel the vibrations of an angler's footsteps when he walks on the bank of a pool or on nearby rocks.

Although in most fish the lateral line system is sensitive to weak electrical currents, there are also some that have developed a kind of radar system by transmitting low-voltage electrical pulses ranging anywhere between 50 and 1,600 each second and which are reflected by objects and other fish, to be picked up again by lateral line cells. This is like picking up the echoes of their own sounds. It is not known exactly how these transmissions are made, but the pulse patterns are different in different species, like those of the flashing photophores or luminous organs of many deep-sea fish; they also provide a degree of identification and recognition.

We have no proof that any of the luminous organs used by so many fish are part of the lateral line system, but the system does produce some very bright organs. This can be seen in the mulloway, in which the highly visible "portholes" seem to be part of the lateral line, alternating with nonluminous organs.

The lateral line system, with its great versatility, is probably the most effective organ that nature could have contrived for functioning in dark waters or during the night.

INFRARED ORGANS IN SNAKES

There are snakes with special heat-sensitive organs in their heads that are so effective that some of them will register a change of temperature of as little as a thousandth of a degree centigrade, and the reaction takes no more than a tenth of a

64. The pit vipers (crotalid snakes) have an organ in a pit (*P*), which is sensitive to infrared rays, sensitive, in fact, to anything passing near by that changes the temperature of the air by a fraction of a degree. The rattlesnake (*Crotalus* spp., above) and copperhead (*Agkistrodon mokasen,* below) are both pit vipers. *N* is the nostril, quite distinct from the pit.

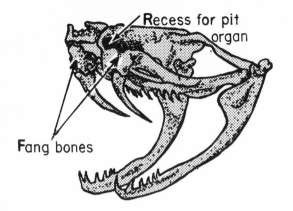

Recess for pit organ

Fang bones

65. In crotalid snakes, the pit organ extends into a recess in the fang bone, which moves to erect the fangs as the mouth is opened.

second. This means that no warm-blooded creature can pass within yards of such snakes without detection, and perhaps attack, by them.

The snakes in which we know most about these organs are the pit vipers (family Crotalidae), and they are distinguished by a large pit on each side of the upper jaw. They can be blinded, or lying around on the darkest night, but will not miss their strike. So far as we know, these snakes have no hearing as we understand it; they sense only vibration through the ground. There are six genera and over eighty species of pit vipers, a few occuring in Asia and Malaysia, and the rest, about fifty species, in America.

Besides all the rattlesnakes, which are typical pit vipers, the group includes the water moccasin (*Agkistrodon piscivorus*), the copperhead (*A. mokasen*), the dreaded bushmaster (*Lachesis muta*), which grows to 12 feet in length, and the fer-de-lance

66. Along the edges of the lower jaws of pythons are rows of deep pits which are sensitive to temperature changes when prey passes, like the facial pits of the crotalids. They can be seen distinctly in the diamond snake (*Morelia* s. *spilotes*) shown here. (Lower): The reticulated python (*Python reticulatus*) shows pits in the front of the upper jaw also.

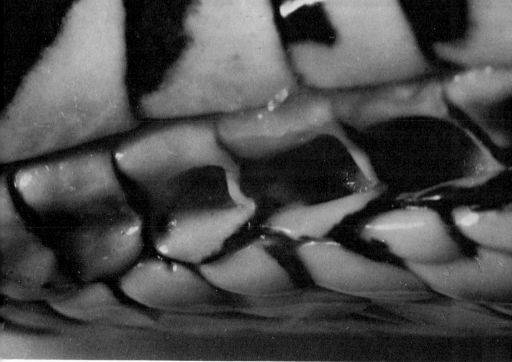

(*Bothrops atrox*). Except for two species, *Atractaspis* and *Causus*, both primitive and crepuscular (active in twilight), their eye pupils are all vertical slits during the daytime, protecting the highly sensitive retinas that go with arhythmic habits. All these snakes, in fact, have excellent night vision, but the infrared organs make it possible for them to attack and capture prey even in complete darkness, when vision of the most sensitive kind will not function.

The pit organ is sensitive to all temperatures and temperature changes above its own temperature, and also to air movement. It expands greatly inside to occupy a large indentation in the movable bone carrying the venomous fangs. It is particularly large in the bushmaster. Within the pit there is a thin membrane stretched between its walls, and it is possible that air in the pit expands as radiation reaches it, so deflecting the membrane slightly, the extent of deflection being related to the amount of rise in temperature. This in turn stimulates the rich nerve supply that connects the pit with the part of the brain responding to sensations in the face and teeth. Although this sensitivity can be measured, the mechanism is not yet fully understood, for such extreme sensitivity to temperature is not known in any other animal.

Experiments with hooded rattlesnakes have proved the effectiveness of the pit organs beyond question. If a balloon containing warm water is brought toward one of them, or passed near it, the snake will strike out at it when it is still some distance away, and will puncture it every time. However, the snake will not react to a balloon containing cold water.

67. The deep pit organs on the lower jaw in Figure 66. (Lower): An even more highly magnified picture that shows their considerable depth.

Although there is no obvious connection between the possession of a thermally sensitive pit organ and the rattling of the tail, it is interesting that only the pit vipers do this. A rattlesnake has a definite structure on the end of its tail for making this warning noise, but some of the other pit vipers, which have no rattles, nevertheless make a creditable effort to imitate one. The copperhead vibrates its tail and creates a rustling sound when it is angry or cornered. The bushmaster acts in the same way, and there have been unconfirmed reports of similar habits in other pit vipers. But the rattlesnake remains the only one with a true rattle.

Other snakes have pits that are thermally sensitive, but these are in different locations. Pythons have pits in the scales along the margins of their lower jaws and on the front of the head, and these seem to react in exactly the same way as a crotalid's pits on the face. These snakes, which lie coiled quietly for long periods, sense the approach of animals with the slight vibration that is conveyed to their bodies, and then locate them precisely with their heat organs.

Perception by animals

Remarkable as all the faculties and senses of nocturnal animals may seem, even more remarkable is the way they learn to use them. How, for instance, does the aye-aye know that the faint sound it hears within a tree trunk is the movement of a good-to-eat insect? How does the dolphin or the seal know that a certain kind of echo means mackerel but another means cod? How does the owl know that one slight sound in the dark means a baby rat but another means a snake too big for it to overpower?

The interpretation of these sounds suggests a high level of perception, such as we use when we touch a smooth surface in the dark and know that it is glass and not polished wood. Perception requires the association of sensation with memory and experience, and although few people will agree to reasoning power in creatures other than man, we do know that some of our own perception is through reasoning in the light of experience. Perhaps animals have a much greater capacity to teach their offspring than we are able to discern.

Teaching by example we can understand; but to what extent is this possible in complete darkness? We cannot ignore the possibility that animals may be able to communicate with each other in much more sophisticated ways than we understand, both by language and sign. A combination of example, communication, and some indefinable quality that we can only think of as instinctive response to basic needs may be all that is involved; but few things are ever quite what they seem to be, and at present the odds are in favor of eventually discovering that animals have unique powers of communication that may teach us a great deal when we begin to understand them.

So far we know that each sense is directly connected to an area in the brain, where its function is analyzed, recorded, and appreciated. Each of these centers subscribe to a system of integration that enables them to influence each other, and to be influenced by memory or experience. Here is a simple example:

A man may be countless miles from civilization, but if he has a map and can associate a single feature of the surrounding scenery with something on that map, he can judge his whereabouts in relation to any other place in the world, even if he has never been there. Individually, the detail on the map and the feature of the landscape would tell him little. It is his ability to connect them in his mind because of past experience or education that enables him to make his decision and therefore to plan his movements.

This would be too much to expect of any other creature, of course, but, in a more limited manner, the higher animals that have evolved what we call perception are able to fix their position in relation to other animals as well as to their surroundings quite efficiently. Such a form of social orientation is very important to animals, but at times it is more complicated than it appears to be; with some animals, gravity, air currents, pressure, the sense of balance, and so on, play a more important part in it than with us.

Vision is undoubtedly the predominant factor in some animals because it is the greatest aid to perception. In others it is hearing; in still others it may be the lateral line system. Integration of these senses goes a step further when an animal has them well enough balanced so that, like the dolphin, it can identify a fish by sound echoes, by vision, by the feel as it captures it, and finally perhaps by the taste. It identifies an object by more than one means, or any of them can be coordinated for the same purpose, such as seeing and hearing at the same time.

Some people accept these powers as perfectly normal in themselves but at the same time fail to realize that in lower animals physical orientation or the ability to judge and fix position in relation to objects and places is just as important as in us because it is essential for immediate survival and must frequently operate twenty-four hours a day. As soon as one of the senses is lost or impaired, animals will fall victim to predators. They keep their perceptual powers in perfect condition or they succumb in the fight for survival.

For animals active in the hours of darkness, the struggle to survive may be even greater than for those active in the daytime. There are often more problems finding mates; it is harder for the young to learn from adult example; though detection of food may be highly efficient, it may often be more difficult. It shows the remarkable adaptation of their senses, therefore, that most animals active at night are able to survive to adulthood and to propagate their species.

Index

(Figures in italics refer to illustrations.)

lamprey, *124*
lantern fish, *54*
lateral line system, *121*, 121, *124*, 126
Lepisosteus osseus, 123
Lialis burtonis, 18
lions, *15*
lizard, *18*, 26
  burrowing, 26
  giant prehistoric, *117*
luciferin, *58*, 59
luminous bacteria, 60, 61
luminous clouds, 61
luminous fluids, 61
luminous insects, 61, 63
luminous organs, *25*, 53, *54*, 55, 57, *58*, 59, 128
luminous shrimp, 61

Macroderma gigas, 75, 80
Macrotus, 81, *84*
mammals, 10
  marsupial, 10
  monotreme, 10
  placental, 10
marsupials, 10
Megalania prisca, 117
mole, 60, 120
  star-nosed, 120
monotremes, 10
moray eel, *107*
Mordacia mordax, 124
Morelia s. *spilotes, 131*

mouse, hopping, *68*
mulloway, *124*
Muraenidae, *107*
Mustelus canis, 45
Myrmecophaga tridactyla, 109, 109

night vision, 27
Notomys mitchelli, 68
Nyctophilus geoffroyi, 75

ocean depth zones, *55*
octopus, *42*
Octopus cyaneus, 42
oilbird, 86
opossum, Virginia, *67*
Orcinus orca, 95
Orectolobus maculatus, 45, 113
organs, infrared, 128
Orycteropus afer, 66, 100
Otocyon megalotis, 66, 100
owl, barn, *41*

Papuan frogmouth, *41*
Philippine tarsier, *69*
photophores, *54*, 55, *58*, 128
Pilodictus, 113
pinna, *64*, *66*, *67*, *68*, 70, 73
pit organs, 128, *129*, *130*, *131*, *132*, 133
pit vipers, *129*, *130*, 130
Plecotus auritus, 76
Podargus papuensis, 41